BIOTECHNOLOGY IN A ... AND MEDICINE

UNIQUE ENZYMES OF *ASPERGILLUS* FUNGI USED IN JAPANESE BIOINDUSTRIES

BIOTECHNOLOGY IN AGRICULTURE, INDUSTRY AND MEDICINE

Additional books in this series can be found on Nova's website under the Series tab.

Additional E-books in this series can be found on Nova's website under the E-books tab.

BIOTECHNOLOGY IN AGRICULTURE, INDUSTRY AND MEDICINE

UNIQUE ENZYMES OF *ASPERGILLUS* FUNGI USED IN JAPANESE BIOINDUSTRIES

EIJI ICHISHIMA

Nova Science Publishers, Inc.
New York

For permission to use material from this book please contact us:
Telephone 631-231-7269; Fax 631-231-8175
Web Site: http://www.novapublishers.com

NOTICE TO THE READER

LIBRARY OF CONGRESS CATALOGING-IN-PUBLICATION DATA

Unique enzymes of Aspergillus fungi used in Japanese bioindustries / editor, Eiji Ichishima.
p. cm.
Includes bibliographical references and index.
ISBN 978-1-61209-719-0 (hardcover)
1. Enzymes--Synthesis. 2. Aspergillus--Industrial applications. 3. Biotechnology industries--Japan. I. Ichishima, Eiji, 1934-
TP248.E5U55 2011
572'.7--dc23

2011031513

Published by Nova Science Publishers, Inc. † New York

CONTENTS

PREFACE

The most widely used organisms are fungi, and several enzymes and organic acids are synthesized by species of *Aspergillus*. Over the past 1,000 years, the use of hydrolytic enzymes from fungi has become more prevalent in Japanese fermentation industries. The molds *Aspergillus oryzae*, *A. sojae*, *A. awamori* and *A. saitoi* are of great practical importance in fermentation industries, enzyme technologies and food industries in Japan. This new book discusses the catalytic and molecular properties of unique and characteristic enzymes obtained from *Aspergillus* fungi used in Japanese bioindustries.

Chapter 1

INTRODUCTION

The most widely used organisms are fungi, and several enzymes and organic acids are synthesized by species of *Aspergillus*. Over the the past 1,000 years, the use of hydrolytic enzymes from fungi has become more prevalent in Japanese fermentation industries. The molds *Aspergillus oryzae* (Figure 1), *A. sojae*, *A. awamori*, and *A. saitoi* are of great practical importance in fermentation industries, enzyme technologies, food industries, and civilization in Japan [1]. In the eastern world rice is used instead of malt or mashed grapes for fermentation. Here, a mold, usually *A. oryzae*, initiates the fermentation process by hydrolyzing rice starch to fermentable sugars. Later on, the sugar is converted by spontaneous fermentation of either yeast or bacteria, leading to products such as sake, Japanese rice wine [2]. Other examples of the traditional use of fungi in food production are in the making of soy sauce and miso paste. In the fermented vegetable protein, soy sauce, the cooked soybeans are mixed with equal amounts of roasted wheat and then inoculated with a pure cultured Koji starter or seed mold [3].

Another important field for the early industrial application of fungi was the production of enzymes, and enzymes are now being used in a wide variety of processes. The invention and production of Takadiastase for *A. oryzae* by Takamine J. in 1894 became an enzyme industry in the late nineteenth century [4]. After Takadiastase was discovered, it was produced in appreciable amounts during the Second World War, initially with *A. oryzae*, *A. sojae*, *A. awamori*, and *A. niger*. *Aspergillus saitoi*, which is a food microorganism and a black *Aspergillus* used in shochu a traditional Japanese spirit, was described taxonomically by Sakaguchi *et al.* [5]. An acid stable proteolytic enzyme,

Molsin, from *A. saitoi* produced on an industrial scale by us is used in the preparation of a human digestant (Fujisawa Pharmaceutical Co., Osaka) [6, 7].

The genome of *Aspergillus oryzae* was sequenced in 2005 [8]. The ability to secrete large amounts of proteins and the development of a transformation system have facilitated the use of *A. oryzae* in modern biotechnology. The molecular analysis of an inactive aflatoxin biosynthesis gene culture in *A. oryzae* RIB strains was confirmed in 2006 [9]. In 2006, the Brewing Society of Japan gave its seal of approval to the decision: "*Aspergillus oryzae* and related food *Aspergilli* are the national microorganisms of Japan".

In this review, catalytic and molecular properties of unique and characteristic enzymes obtained from *Aspergillus* fungi used in Japanese bioindustries are described.

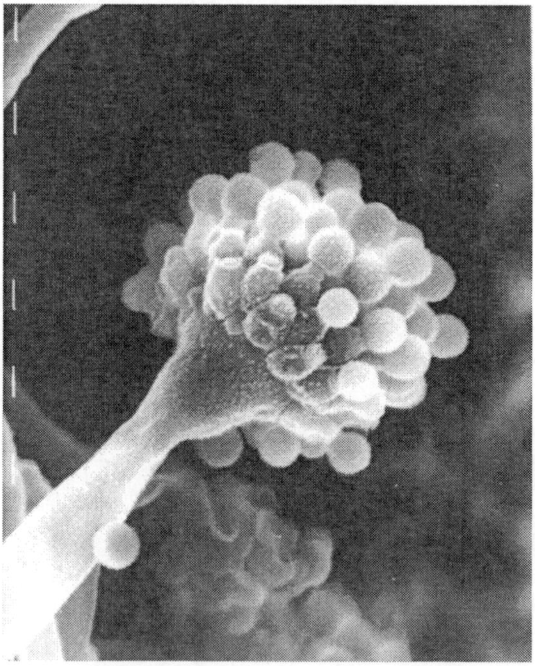

(Reproduced with the kind permission of Dr. Kazuya Hayashi, Kikkoman Corp.)

Figure 1. Electronmicroscopic photograph of conidia, phialides, and conidiophore of the filamentous fungus *Aspergillus oryzae*.

Chapter 2

The Third Active Site Residue of Aspartic Proteinase from *Aspergillus* is Essential for Trypsinogen Activation at Ph 3-4.5

2.1. Aspartic Proteinase

The aspartic proteinases (EC 3.4.23.–) from mammals and fungi are all characterized by a molecular mass of 35,000, and pH optimal for catalytic action in the range of 1.5-5.0. They include the mammalian enzymes pepsin (EC 3.4.23.1), cathepsin D (EC 3.4.23.5), renin (EC 3.4.23.15), and chymosin (EC 3.4.23.4), and a number of fungal enzymes such as those from *Aspergillus saitoi* , *A. oryzae* and *A. sojae* (EC 3.4.23.18), *Penicillium janthienllum* (EC 3.4.23.20), *Rhizopus chinensis* (EC 3.4.23.21), and *Rhizomucor* (formerly *Mucor*) *pusillus* (EC 3.4.23.23), *Endothia parasitica* (EC 3.4.23.22), and a viral enzyme from human immunodeficiency virus (HIV-I) (retropepsin, EC 3.4.23.16). They share the property of being inhibited by an epoxy-compound, 1,2-epoxy-3-(*p*-nitrophenoxy)-propanne (EPNP), and by a diazonium compound, diazoacetyl-DL-norleucine-methylester (DAN), in the presence of Cu^{2+} ions.

Aspartic proteinases comprise a group of enzymes whose proteolytic activities are dependent on two aspartic acid residues, Asp32 and Asp215, in pepsin numbering [10]. There is good evidence that most of them belong to a family of enzymes similar in sequence and structure. These enzymes are bilobal, each lobe having a similar fold, with a deep and extended cleft that can accommodate at least seven amino acid residues of substrate in the S_4-S_3' subsites [11]. The abbreviations used are: S_1, S_2, S_3, etc. and S_1', S_2', S_3', etc., for corresponding subsites of the proteinase [12]; P_1, P_2, P_3, etc. and P_1', P_2', P_3', etc., for amino acid residues of substrate on the amino-terminal and carboxyl sides of the scissile peptide bond, respectively. Each domain contributes one aspartyl residue, corresponding to Asp32 and Asp215 of pepsin, to the catalytic center of the extended binding site [13].

The flap, an antipararell β-hairpin comprising residues 73 to 83 [penicillopepsin [14] numbering], projects across the cleft, forming a channel into which a substrate binds. Although the enzymes are quite similar in their three-dimensional structures, there are drastic differences in the catalytic properties, especially in substrate specificities. Meanwhile, these enzymes show distinct preferences for cleaving the bonds between hydrophobic residues occupying the P_1-P_1' subsites. For example, renin [15] is characterized by very specific hydrolysis of angiotensin at Leu10-Leu11 and chymosin [16] digests κ-casein at Phe105-Met106 for milk clotting.

Thus, the differences in substrate specificity may be the result of alteration in the structure of substrate binding sites.

2.2. Aspergillopepsin I

An aspartic proteinase from *A. saitoi* ATCC 14332 (now designated *A. phoenicis* [17]), aspergillopepsin I (formerly aspergillopeptidase A), is a monomeric protein with a molecular mass of 34,302 Da [7, 8]. Aspergillopepsin I primarily hydrolyzes two bonds in the oxidized B chain of insulin: Leu15-Tyr16 and Phe24-Phe25 [18]. Minor cleavages are noted at His10-Leu11, Ala14-Leu15 and Tyr16-Leu17.

Aspergillopepsin I favors hydrophobic residues at P_1 and P_1' [18, 19] but also accepts a Lys residue in the P_1 position, which leads to activation of trypsinogen by cleavage of the Lys6-Ile7 bond [20, 21]. Furthermore, penicillopepsin also has a preference for a Lys residue in the P_1 position, which leads to activation of trypsinogen by cleavage of the Lys6-Ile7 bond [22]. The S_1 subsite is formed by several hydrophobic residues in the neighborhood of the catalytic Asp-32 and by the residues on the active site flap.

For trypsinogen activation, enteropeptidase (enterokinase, EC 3.4.21.9) [23] is a key enzyme for mammalian protein digestion. The selective cleavage site of trypsinogen by enteropeptidase initiates Lys6-Ile7 bond. Then trypsin

(EC 3.4.21.4) activates other zymogens. Thus, the formation of trypsin by enteropeptidase is the master activation step.

Table 1. Comparison of the amino acid sequence of the active-site flap in the aspartic proteinase family

Aspartic proteinase	Residue number 70 83
Class I	
Aspergillopepsin I	WDISYG**D**G**S**SASGD
Penicillopepsin	WSISYG**D**G**S**SASGN
Rhizopuspepsin	WSISYG**D**G**S**SASGI
Endothiapepsin	WSISYG**D**G**S**SSSGD
Candidapepsin	FYIGYG**D**G**S**SSQGT
Class II	
Mucorpepsin	LNITYG**T**G-GANGI
Porcine pepsin	LSITYG**T**G-SMTGI
Human cathepsin D	FDIHYG**S**G-SLSGY

(Shintani,T. *et al., J. Biochem. (Tokyo), 120*, 974-981 (1996)).

The amino acid sequences of aspergillopepsin I, penicillopepsin, rhizopuspepsin, endothiapepsin, candidapepsin, mucorpepsin, porcine pepsin, and human cathepsin D are aligned. These enzymes are classified into two groups (class I and II). Class I enzymes have the ability to activate trypsinogen, and class II enzymes have no activity for trypsinogen activation. The mutated positions are indicated in boldface characters.

Aspartic proteinases generally have broad primary specificity, but the fungal and mammalian enzymes can differ dramatically from each other in catalytic efficiency against substrates containing a Lys residue at the P_1 position [24]. Based on the sequence alignment and structure of aspartic proteinases, the major differences in the S_1 subsite between the fungal and mammalian enzymes can be located at position 76 [77, where the number in square brackets denotes penicillopepsin numbering] and 78 [79] on the active site flap (Table 1). These residues may be involved in the recognition of basic amino acid residues at the P_1 position of substrates.

Aspergillopepsin I and the fungal aspartic proteinases, penicilllopepsin, rhizopuspepsin, and endothiapepsin, display significant similarities in their amino acid sequences. Although no three-dimentional structure of aspergillopepsin I is available, the substrate binding sites may be identical by analogy with those aspartic proteinases, penicillopepsin [14, 25, 26], endothiapepsin [27], and rhizopuspepsin [28].

2.3. D76S-Aspergillopepsin I Abolished Trypsinogen Activation

We isolated and characterized the genomic DNA (D25318) and cDNA of the aspergillopepsin I gene (*apnS*) and expressed it in yeast *S. cerevisiae* cells [29]. By altering a single substituted amino acid in aspergillopepsin I by site-directed mutagenesis to the corresponding residue in pepsin, renin, or chymosin, we obtained a mutant enzyme denoted D76S-aspergillopepsin I [29] in Figure 2. These mammalian aspartyl proteinases have no capacity of trypsinogen activation and have threonine residue instead of Asp76 in aspergillopepsin I. Expression products of mutant D76S-aspergillopepsin I and wild-type enzyme were observed by SDS-PAGE and Western blotting. These results strongly suggest that active D76S-aspergillopepsin I was secreted from the yeast cells harboring a plasmid designated to express prepro-D76S-aspergillopepsin I in the cells. The cultured filtrates of recombinant wild-type and mutant enzymes were isolated and were salted out with 60% $(NH_4)_2SO_4$. Then, the precipitates were dialyzed against 10 mM acetate buffer (pH 5.0). The activities of the two recombinant enzymes for casein hydrolysis [7] were 0.35 and 0.37 nkatal/ml, respectively. After activation of trypsinogen at pH 3.0, the fluorogenic activity of activated trypsin was measured at pH 8.0 with benzoyl-L-arginyl-4-methylcoumaryl-7-amide (Bz-Arg-MCA). Trypsinogen activating activity of the wild-type enzyme was determined to be $(1.08 \pm 0.03) \cdot 10^{-2}$ nkatal/ml, while none was found for the mutant enzyme.

To explore the contribution of an unpaired charged residue to substrate binding and catalysis in the binding pocket of D76S-aspergillopepsin I, we were interested in how interaction of the ε-amino group of lysyl substrate, trypsinogen, affects the catalytic efficiency. As shown in Figure 2, the activity of D76S-aspergillopepsin I on trypsinogen was completely abolished. Hydrolysis of casein sustrate by the D76S-aspergillopepsin I mutant is not altered within this pH range. This suggests that the activity is quantitatively influenced by the electrostatic interaction between ε-amino group in the basic substrate of trypsinogen and acidic Asp76 in aspergillopepsin I.

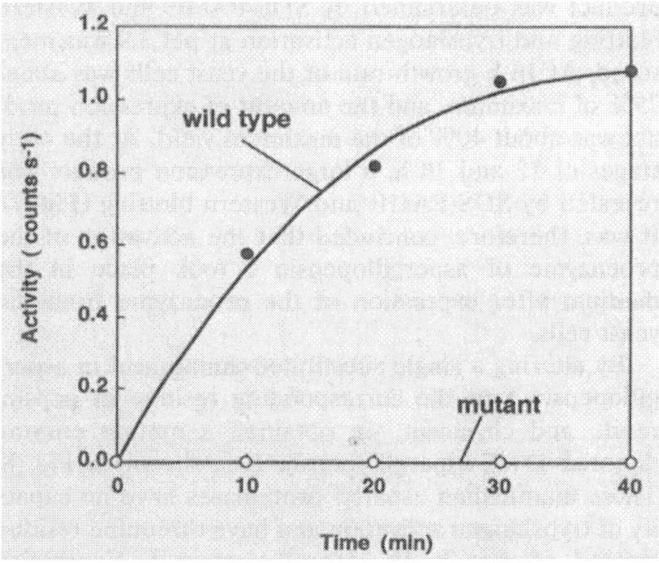

Figure 2. Activities of trypsinogen activation of the mutated D76S-aspergillopepsin I and aspergillopepsin I. After activation of trypsinogen at pH 3.0, the fluorogenic activity of activated trypsin was measured at pH 8.0 with Bz-Arg-MCA.

(Shintani, T. and Ichishima, E., *Biochim. Biophys. Acta, 1204*, 257-264 (1994))

Thus, we concluded that Asp76 is the binding site for the basic substrate, trypsinogen, and that a single amino acid mutation (Asp76 → Ser76) is sufficient to alter the substrate specificity of the enzyme for trypsinogen activation [29].

2.4. Characterization of the S_1 Subsite Specificity

In order to obtain further information about the primary specificity of aspergillopepsin I, we constructed an aspergillopepsin I expression system in *Escherichia coli*, prepared mutants at positions 76 and 78 of aspergillopepsin I by site-directed mutagenesis, and compared their molecular and enzymatic properties [30)] We showed that the nature of the active site flap is important for deciding the S_1 substrate specificity of aspartic proteinases [30].

For overexpression, the ATG codon was added in front of the putative pro-sequence of aspergillopepsin I cDNA by the PCR technique. The cDNA fragment was cloned into pET12a, and the resulting plasmid, pETAP was introduced into *E. coli* BL21 (DE3) cells. The cells harboring pETAP produced pro-aspergillopepsin I under the control of T7 promoter, and the enzyme was not active and was mostly insoluble. We therefore developed a procedure for the renaturation of recombinant pro-aspergillopepsin I into the correctly refolded zymogen, as described in "EXPERIMENTAL PROCEDURES" of the references [30]. The inclusion bodies were solubilized with 8 M urea, and dialyzed under alkaline pH. The active proenzyme was purified to homogeneity by two chromatography steps on a Toyopearl HW-55F column and a RESOURCE Q column. The purity of protein was judged by SDS-PAGE (Figure 3, lane 1). The yields of purified pro-aspergillopepsin I from inclusion bodies were routinely 10-20%. The amino acid sequence of the NH_2-terminal region was determined as Met-Ala-Pro-Ala-Pro-Thr-Arg-Lys-Gly-, identical to the expected from the construction. Pro-aspergillopepsin I was converted to the mature form within 5 min by incubating at pH 2.7, 0°C (Figure 3, lane 3).

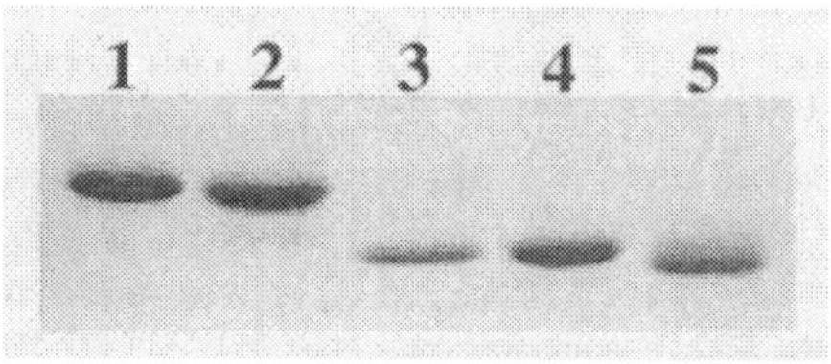

Figure 3. Conversion of proaspergilopepsin I to mature aspergillopepsin I. Purified proaspergillopepsin I was incubated in 0.2 M glycine-HCl, pH 2.7, in the presence (lane 2) or absence (lane 3) of 0.1 mM pepstatin A at 25°C for 30 min. After the incubation, samples were subjected to SDS/PAGE on a 10% gel. The gel was stained with Coomassie Brilliant Blue R-250. Lane 1, purified proaspergillopepsin I; lane 4, purified mature aspergillopepsin I; lane 5, native aspergillopepsin I.

(Shintani, T. *et al., J. Biochem. (Tokyo), 120*, 978-981 (1996))

The conversion was inhibited by pepstatin, an aspartic proteinase inhibitor (Figure 3, lane 2). These results indicate that aspergillopepsin I was generated from pro-aspergillopepsin I by autoactivation.

Purified aspergillopepsin I was prepared from the crude pro-aspergillopepsin I preparation by acidification and subsequent chromatography on a RESOURCE Q column, and gave a single band on SDS-PAGE (Figure 3, lane 4). The NH_2-terminal sequence of recombinant aspergillopepsin I was found to be Glu-Ala-Ala-Ser-Lys-Gly-Ser-Ala-, which was three amino acid residues longer than that of native aspergillopepsin I from *Aspergillus saitoi* (Ser-Lys-Gly-Ser-Ala-). These results suggested that autocatalytic cleavage of the propeptide (Lys-Glu) may be occurring in front of glutamic acid and the product is terminated by aminopeptidase(s) in the fungal culture or that the activation of the zymogen *in vitro* may be basically different from that *in vivo*. In spite of the difference of NH_2-terminal sequences of the authentic and recombinant aspergillopepsin Is, these two enzymes were almost equal in specific activity for hydrolysis of casein (0.138 and 0.139 katal/kg, respectively) and kinetic parameters for hydrolysis of a peptide substrate, Pro-Thr-Glu-Phe-Phe(4-NO_2)-Arg-Leu, at pH 3.0 (K_m = 0.01 mM and k_{cat} = 15.5 s^{-1} for both enzymes).

The mutant aspergillopepsin Is were purified by the same method as the wild-type aspergillopepsin I. The purities of the enzymes were judged by SDS-PSGE. The mobilities of mutant aspergillopepsin Is were the same as that of the wild-type enzyme from *E. coli*. Furthermore, the effects of the maturation on the secondary structures of the enzymes were investigated by CD. The CD spectra (from 190 to 260 nm) of the enzymes were essentially identical, and contents of α-helix and β-structure of the wild-type and mutant enzymes were very similar. These results indicated that there was no significant conformational difference between the wild-type and mutant aspergillopepsin Is.

Table 2 shows a comparison of the specific activities of wild-type and mutant aspergillopepsin Is for the hydrolyses of casein, hemoglobin and the activation of bovine trypsinogen by cleaving the Lys6-Ile7 bond. The activity of trypsinogen activation fell below the detectable level in the D76S-mutant without any change in the proteinase activities for casein and hemoglobin. However, in the D76N-mutant both hydrolytic and trypsinogen activating activities were approximately half those of wild-type. In the D76E and S78A mutants there were significant decreases in the activity of trypsinogen activation, with very little effect on the proteinase activities for casein and hemoglobin. The catalytic activities of D76T and ΔS78 (deletion mutant) were generally decreased, but there were larger reductions in trypsinogen activating activity, as with D76S- and S78A-aspergillopepsin Is. These results strongly suggest that Asp76 [Asp77 in penicillopepsin numbering] and Ser78 of aspergillopepsin I play an important role in the recognition of the basic amino acid residue at the P_1 position.

The wild-type [19] and D76N-aspergillopepsin Is cleaved the Tyr4-Ile5 and His6-Pro7 bonds of angiotensin I (Asp-Arg-Val-Tyr-Ile-His-Pro-Phe-His-Leu), while only the Tyr4-Ile5 bond was hydrolyzed by D76E, D76S, D76T, S78A, and ΔS78 aspergillopepsin Is. Thus, the mutations of Asp76 and Ser78 caused changes in the substrate specificities of aspergillopepsin I, decreasing the affinity for basic residues at the P_1 position and making the specificity more restrictive so that only hydrophobic residues are acceptable at the P_1 and P_1' position, as in mammalian aspartic proteinases.

Table 2. Specific activities of wild-type and mutant aspergillopepsin Is for the hydrolyses of casein and hemoglobin and the activation of bovine trypinogen

Proteolytic activities were assayed in 50 mM sodium acetate, pH 2.7, 1% casein or hemoglobin at 30°C. The activation of bovine trypsinogen was done with 5 mM trypsinogen in 50 mM sodium citrate, pH 3.0 at 30°C. Parentheses indicate the percentages of the activities of wild-type aspergillopepsin I.

Mutant aspergillo-pepsin I	Specific activity		
	Casein (katal/kg)	Hemoglobin (katal/kg)	Trypsinogen activation (katal/kg)
Wild type	0.139 (100)	0.75 (100)	5.1 (100)
D76N	0.060 (43)	0.38 (51)	2.3 (45)
D76E	0.041 (29)	0.45 (60)	0.02 (0.4)
D76S	0.103 (74)	0.40 (53)	N.D.
D76T	0.023 (17)	0.12 (16)	N.D.
S78A	0.142 (102)	0.53 (71)	0.37 (7)
ΔS78	0.010 (7)	0.077(10)	N.D.

(Shintani,T. *et al., J. Biochem. (Tokyo), 120*, 974-981 (1996))

To probe the effects of the mutations on specificity at the P_1 position, we used peptide substrates, Pro-Thr-Glu-Phe-Phe(4-NO_2)-Arg-Leu (PTEFF(4-NO_2)RL, 'peptide A'), Pro-Thr-Glu-Lys-Phe(4-NO_2)-Arg-Leu (PTEKF(4-NO_2)RL, 'peptide B'), and Ac-Ala-Ala-Lys-Phe(4-NO_2)-Ala-Ala-amide (Ac-AAKF(4-NO_2)AA-NH_2, 'peptide C'). Peptide A was cleaved at the Phe-Phe(4-NO_2) bond, which represents pepsin-like activity, and peptide B and peptide C were cleaved at the Lys-Phe(4-NO_2) bond, which is equivalent to trypsinogen activation. No other bond in these peptides was hydrolyzed by any of the enzymes. From kinetic determination of the k_{cat}/K_m (catalytic efficiency) one can obtain the second-order rate constant for conversion of substrate to product. Combining the effects due to substrate binding and transition state stabilization, this parameter is useful for assessing altered substrate specificity. Differences in $\log(k_{cat}/K_m)$ provide an accurate measure of the lowering of the transition state activation energy ($\Delta G_T^{\neq}$). The k_{cat}, K_m, and k_{cat}/K_m values were determined from the initial rate measurements and the results are listed in Table 3 and graphically represented in Figure 4.

The effects of individual mutations on the activity of aspergillopepsin I towards the substrate that contain Phe at P_1 were considerably less dramatic. The mutations at Asp76 and Ser78 led to a 2- to 30-fold lowering of k_{cat}/K_m compared to the wild-type aspergillopepsin I for this substrate. The decrease in k_{cat}/K_m was largely due to the effect of the mutation on K_m.

Reactions were carried out in 0.1 M sodium acetate, pH 4.5 at 30°C. The kinetic parameters were determined from plots of initial rates *versus* substrate concentration (0.002-0.300 mM for all substrates).

Table 3. Kinetic parameters for wild-type and mutant aspergillopepsin Is towards the peptide substrates containing phenylalanine or lysine residue at the P_1 position

Mutant aspergillopepsin I	K_m (mM)	k_{cat} (s^{-1})	k_{cat}/K_m ($mM^{-1} \cdot s^{-1}$)
PTEF↓F(4-NO_2)RL			
Wild type	0.007	15.7	2,240
D76N	0.010	8.3	830
D76E	0.057	10.9	191
D76S	0.061	10.8	177
D76T	0.073	5.9	81
S78A	0.017	20.3	1,190
ΔS78	0.107	10.8	101
PTEK↓F(4-NO_2)RL			
Wild type	0.009	4.5	500
D76N	0.027	3.2	119
D76E	0.050	0.20	4
D76S	0.034	0.040	1.2
D76T	0.065	0.014	0.2
S78A	0.017	0.61	36
ΔS78	0.046	0.036	0.8
Ac-AAK↓F(4-NO_2)AA-NH_2			
Wild type	0.092	6.8	74
D76N	0.103	3.5	34
D76E	0.171	0.039	0.2
D76S		ND	
D76T		ND	
S78A	0.130	0.93	7
ΔS78	0.094	0.001	0.01

(Shintani,T. *et al., J. Biochem. (Tokyo), 120*, 974-981 (1996))

Arrows (↓) indicate the scissible bonds.

ND, not ditected.

There was a significant effect of the mutations on activity towards the substrate containing Lys at P_1. For peptide B, the D76E, D76S, D76T, and ΔS78 mutations caused 130-, 430-, 2,600-, and 650-fold decreases in k_{cat}/K_m value compared to the wild-type aspergillopepsin I, respectively, whereas the D76N mutant effectively hydrolyzed this substrate. Peptide B has the same sequence except for the substitution of P_1 Phe to Lys. The effect of each mutation on K_m was similar to that in the case of peptide A, indicating that the changes of K_m value caused by the mutation may be independent of the interaction with the P_1 side chain. The decreases in catalytic efficiencies mainly reflect decreases in k_{cat} for the Lys side chain of the substrate.

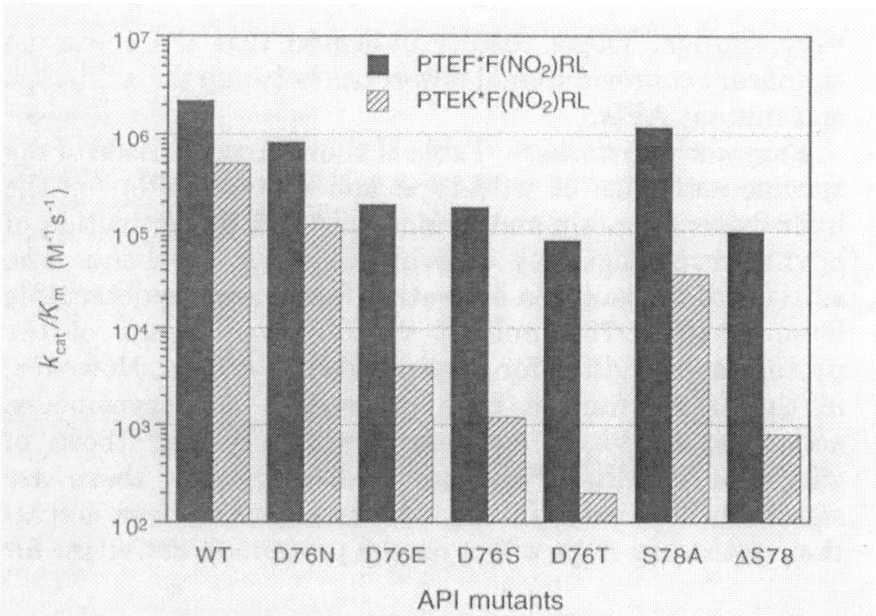

Figure 4. Catalytic efficiencies (k_{cat}/K_m) for mutant aspergillopepsin I against peptide substrates having Phe or Lys at the P_1 positions.

(Shintani, T. *et al., J. Biochem. (Tokyo), 120*, 974-981 (1996))

The effects of each mutation on activity towards peptide C were similar to those in the case of peptide B, although there was no detectable cleavage with D76S and D76T mutant aspergillopepsin Is.

We have mutated the Asp76 [77, penicillopepsin numbering] and Ser78 [79] residues of aspergillopepsin I to test their roles in determining the primary specificity. These residues are conserved only in the enzymes with the ability to activate trypsinogen, such as penicillopepsin and rhizopuspepsin. The three-dimensional enzyme-inhibitor complex structutres have been analyzed for penicillopepsin and rhizopuspepsin. According to these analyses, the side-chain carboxylate and amide nitrogen of Asp77 are hydrogen bonded to P_2 amide nitrogen and carbonyl of the inhibitor, respectively, and there is a hydrogen bond between the P_2 carbonyl oxygen and amide nitrogen of Gly76 (Figure 5A). When a hydrophobic amino acid occupies the P_1 position of a ligand, the hydrophobic side chain is accommodated in the hydrophobic S_1 pocket, which consists of Asp31, Tyr75, Phe112, and Leu121 [26, 27]. The three-dimentional structure of the penicillopepsin/Iva-Val-Val-LySta-OEt (Iva: isovaleryl; LySta: 4*S*,3*S*-4,8-diamino-3-hydroxyloctanoic acid) complex revealed that the ε-amino group of the LySta side chain forms a tight ion pair with the carboxylate of Asp77 and hydrogen bonds with the carboxylate of Asp77 and hydroxyl group of Ser79 (Figure 5A). We therefore altered Asp76 of aspergillopepsin I to Asn in order to replace the carboxyl group with an uncharged side chain very similar in size to that of Asp and to Glu so that the carboxyl group would be retained but the effective distance to the side chain of the P_1 residue of a substrate would change. The substitution of Ser78 to Ala was made to examine the effect of hydrogen bonding on P_1 Lys specificity by the removal of the hydroxyl group of the Ser side chain.

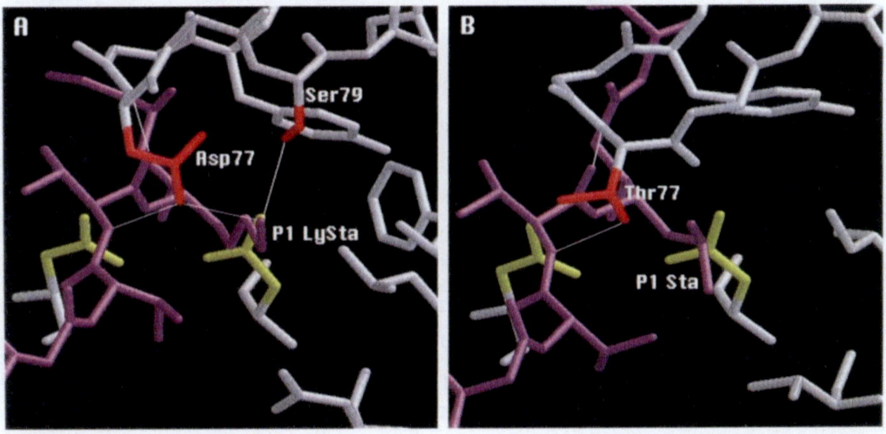

Figure 5. Representations of S_1 subsites of penicillopepsin (A) and porcine pepsin (B) with inhibitors. The crystal structure of these enzymes were retrieved from Brookhaven Protein Data Bank (PDB).

(Shintani, T. *et al., J. Biochem. (Tokyo), 120*, 974-981 (1996))

The D76S and D76T substituted and Ser78 deleted (ΔS78) aspergillopepsin Is corresponding to mammalian aspartic proteinases were also constructed. The deletion of Ser78 (ΔS78) in aspergillopepsin I, corresponding to mammalian aspartic proteinases such as pepsin (Figure 5B), renin, and cathepsin D, would give rise to a shorter and warped flap, which may change the direction of the Asp76 side chain binding to the main chain of the substrate and/or cause steric hindrance between the resulting flap and the ε–amino group of P_1 Lys. Thus, in the recognition of P_1 Lys of substrate, Asp76 [Asp77 in penicillopepsin numbering] and Ser78 of aspergillopepsin I appear to contribute to the stabilization of the transition-state complex by forming a hydrogen bond network between P_1 Lys and the side chain of Asp76 and Ser78, which may be supported by the unique flap structure of aspergillopepsin I.

Furthermore, Asp77 is shown to be the binding site to P_1 Lys in a substrate by crystallographic study of penicillopepsin (25, 26) and site-directed mutagenesis studies of rhizopuspepsin [31]. Using the *S. cerevisiae* expression system, Tyr77 on the flap of mucorpepsin [32] has been shown to enhance catalytic efficiency of this fungal aspartic proteinase.

A comparison of the Michaelis-Menten kinetic parameters for the hydrolysis of peptide A and B by the site directed mutants of aspergillopepsin Is suggests that Asp76 and Ser78 play important roles in aspergillopepsin I activity when substrates contain Lys at the P_1 position. The inhibition analysis of D77T mutated rhizopuspepsin using Lys-Pro-Ala-Ala-[X]-Ala-Leu-Gly-NH2, where [X] is LySta (4*S*,3*S*-4,8-diamino-3-hydroxyloctanoic acid) or As-LysSta, showed that rhizopuspepsin is able to bind substrates containing Lys at the P_1 position through the electrostatic interaction of Asp77 with P_1 Lys [31]. However, D76N mutant aspergillopepsin I retained sufficient activities towards peptides B and C containing Lys at the P_1 position, as well as trypsinogen activation, which indicated that that the negative charge of Asp76 may be dispensable for the recognition of P_1 Lys. On the basis of the data in Table 3, although the changes of K_m value by the mutations may be independent of the interaction with the P_1 side chain, the D76S, D76T, and Δ78 mutations result in large reductions in k_{cat} value *versus* the substrates containing Lys at the P_1 position. It appears that these residues on the active site flap have important roles in the stabilization of the transition state complex. This effect can be explained simply by one or both of the mechanisms: (i) a disturbance of the hydrogen bond network between P_1 Lys and the side chains of Asp76 and Ser78, and/or (ii) steric hindrance of substrate binding resulting from the change of flap structure. When D76S or

D76T mutant aspergillopepsin I binds to a substrate containing Lys at the P_1 position, the hydroxyl group of Ser or Thr would be hydrogen bonded to P_2 amide nitrogen of the main chain as in mammalian enzyme (Figure 5 B) [33, 34], which may cause an alteration of the side-chain torsion angle of Ser or Thr in order to hydrogen bond to the substrate and disable this group from binding to the ε-amino group of P_1 Lys. The substitution of Ser-78 [79] to alanine removes the hydrogen bond between Ser78 and the ε-amino group of P_1 Lys. The deletion of Ser78 in aspergillopepsin I, corresponding to mammalian aspartic proteinases such as pepsin (Figure 5 B), renin and cathepsin D, would give rise to a shorter and warped flap, which may change the direction to the Asp76 side chain to bind to the main chain of the substrate and/or cause steric hindrance between the resulting flap and the ε-amino group of P_1 Lys. Thus, in the recognition of P_1 Lys of a substrate, Asp76 and Ser78 of aspergillopepsin I appear to contribute to the stabilization of the transition-state complex by forming a hydrogen bond network between P_1 Lys and the side chains of Asp76 and Ser78, which may be supported by the unique flap structure of aspergillopepsin I.

Chapter 3

ENGINEERING OF PORCINE PEPSIN

3.1. PEPSIN

Pepsin (EC 3.4.23.1) is a typical aspartic proteinase produced in the gastric mucosa of vertebrates as a zymogen form [10]. This enzyme has been extensively characterized, and its three-dimensional structure has been determined at high resolution. Porcine pepsin, in particular, has been studied as a model to analyze the structure-function relationship of the aspartic proteinases. Although the aspartic proteinases including mammalian and fungal enzymes are quite similar in their three-dimentional structures, there are drastic differences in the catalytic properties, especially in substrate specificities.

There also have been many studies on the tertiary structures of aspartic proteinases with peptide-derived inhibitors, and the relationship between their structures and substrate specificities has been demonstrated. Although the hydrogen bonding pattern between the main chain of inhibitor and enzyme is well conserved in all inhibitor-enzyme complexes, the differences in the size of substrates, of which residues make van der Waals contacts with the side chain of inhibitor, may control the substrate specificities in aspartic proteinases. The attempts to examine the structural determinations of substrate specificities of aspergillopepsin I by site-directed mutagenesis have been made [29, 30], and Asp77 is shown to be the binding site to P_1 Lys in a substrate by crystallographic study of penicillopepsin [25, 26].

The present chapter described the investigation of whether the active site flap controls the S_1 subsite specificity of aspartic proteinases, and for this we generated several mutants of porcine pepsin by site-directed mutagenesis and analyzed their enzymatic properties [35]. The substitution of Thr77 by Asp and the insertion of Ser between Gly78 and Ser79, corresponding to the fungal enzymes, conferred upon porcine pepsin the ability to hydrolyze the substrate containing a Lys residue at the P_1 position. These experiments provide positive evidence of the importance of the active site flap of aspartic proteinases in the S_1 subsite specificity.

3.2. Alteration of S_1 Substrate Specificity of Pepsin to Those of Fungal Aspartic Proteinase

The structure of porcine pepsinogen cDNA was reported earlier by Lin *et al.* [36]. Therefore, porcine pepsinogen cDNA was cloned by reverse transcriptase-PCR, using total RNA from porcine gastric mucosa as a template. PCR primers were designated for direct expression in the *E. coli* pET system. The *Nde*I/*Sal*I fragment was inserted into pET12a, which was designated as pETPP, to express porsine pepsinogen under the control of a T7 promoter. The amplified cDNA was dideoxy-sequenced to verify whether it was synthesized correctly. The sequence was identical to that reported by Lin *et al.* [36] except for the substitution of Tyr-242 by Asp. Porcine pepsin has been sequenced in several laboratories by protein and cDNA sequencing [37, 38], and these studies indicated that there are two variants at position 242 of the amino acid sequence (Asp or Tyr). The electrospray mass spectrometry of porcine pepsin also demonstrated the existence of two variants [39]. Hence, the cDNA obtained in this study seemed to be correct and available for expression studies.

The *E. coli* cells harboring pETPP produced pepsinogen as an inclusion body under the control of the T7 promoter. The inclusion body was purified by washing with a buffer and then refolded by solubilization with 8 M urea and subsequent dialysis under alkaline pH as described in "Experimental Procedures" of references [35]. Porcine pepsin was purified from the refolded pepsinogen preparation by acidification and subsequent chromatogaraphy on a RESOURCE Q column and gave a single band on SDS-PAGE. The NH_2-terminal sequence of recombinant pepsin was found to be predominatly Ile-Gly-Asp-Glu-Pro-Leu-Glu-Asn-, which is known as the NH_2-terminal sequence of porcine pepsin [37], and a minor sequence of Ala-Ala-Leu-Ile-Gly-Asp-Glu-Pro- was also detected. The same results were reported in the activation of native porcine pepsinogen by Kageyama and Takahashi [40]. The secondary structure of native and recombinant porcine pepsins was analyzed by CD spectrometry. The CD spectral data showed that the spectrum of the recombinant enzyme was essentially superimposable on that of the native enzyme. The native and recombinant enzymes were almost equal in specific activity for hydrolysis of acid-denatured hemoglobin (82 and 83 mili-katals/kg, respectively).

The mutant porcine pepsins, T77D, G78(S)S79, and T77D/G78(S)S79, were purified by the same method as wild-type pepsin, and the purities of the enzymes were judged by SDS-PAGE. The NH_2-terminal sequences of the mutants were the same as that of the wild-type enzyme. The secondary structures of recombinant wild-type and mutant pepsins were analyzed by CD spectrometry to determine whether localized or global changes of structures were induced by the mutations. The CD spectral data showed that the spectra of the mutants were essentially superimposable on that of the wild-type enzyme. These results suggest that no major conformational alterations occurred in the mutant enzymes.

Proteolytic activities for acid-denatured hemoglobin and trypsinogen-activating activities due to limited protelysis of the Lys6-Ile7 bond of trypsinogen were measured with wild-type and mutant pepsins. The specific activities of mutant enzymes, T77D, G78(S)S79, and T77D/G78(S)S79, for hemoglobin hydrolysis were determined to be 53.8, and 50 milli-katals/kg, respectively. Each mutant enzyme effectively hydrolyzed hemoglobin although the activity of G78(S)S79 mutant was 10 times lower than that of the wild-type enzyme. The specific activities of wild-type, T77D, and T77D/G78(S)S79 enzymes for trypsinogen activation were 0.04, 0.01, and 18.5 micro-katals/ml, respectively, and no detectable activity was found for the G78(S)S79 enzyme, which indicates that the activity of trypsinogen activation was significantly increased only by the double mutant. However, it was still 300-fold less than the activity of aspergillopepsin I. SDS-PAGE analysis showed that trypsinogen was converted to trypsin by T77D/G78(S)S79 mutant alone (Figure 6, lane 5) and the cleavage site was found to be Lys6-Ile7 by NH_2-terminal sequencing. These results indicated that the double mutation, T77D/G78(S)S79, in the active site flap altered substrate specificity of porcine pepsin to those of fungal aspartic proteinases, aspergillopepsin I, penicillopepsin, and rhizopuspepsin.

Figure 6. Conversion of trypsinogen to trypsin by mutant pepsins. Bovine trypsinogen (5 μ g) was incubated in 50 mM sodium citrate, pH 3.0, with 160 ng of wild-type (*lane 2*), T77D (*lane 3*), G78(S)S79 (*lane 4*), or T77D/G78(S)S79 pepsin (*lane 5*) at 37°C for 6 h and analyzed by SDS-PAGE on a 12.5% gel. The gel was stained by Coomassie Brilliant Blue R-250. *Lane 1*, trypsinogen in the absence of pepsin.

(Shintani, T. *et al., J. Biol. Chem., 272*, 18855-18861 (1997))

Peptide A (PTEFF(4-NO_2)RL) was cleaved at the Phe-Phe(4-NO_2) bond by each enzyme, which indicated that mutant enzymes also prefer hydrophobic residues in P_1 and P_1' in the substrate. The T77D/G78(S)S79 mutant hydrolyzed the Lys-Phe(4-NO_2) bond in peptide B (PTEKF(4-NO_2)RL) and C (Ac-AAKF(4-NO_2)AA-NH_2), which is equivalent to trypsinogen activation. While, for the wild-type enzyme no hydrolyses at Glu-Lys and Phe(4-NO_2)-Ala bonds in peptide B and peptide C were observed (Table 4). The T77D mutant cleaved at the Lys-Phe(4-NO_2) bond in peptide B, but not at the Phe(4-NO_2)-Ala bond in peptide C. There was no cleavage in peptide B and peptide C by the G78(S)S79 mutant. These results indicate that the T77D/G78(S)S79 mutant acquired the ability to recognize a Lys residue at the P_1 position although the enzyme still exhibited a preference for hydrophobic residues at the P_1 position (Table 4).

From kinetic determination of the k_{cat}/K_m (catalytic efficiency) one can obtain the second-order rate constant for conservation of a substrate to a product. Combining effects due to substrate binding and transition state stabilization, this parameter is useful for assessing altered substrate specificity. Differences in log(k_{cat}/K_m) provide an accurate measure of the lowering of the transition state activation energy ($\Delta G_T^{\neq}$). The kinetic parameters for hydrolysis of Phe-Phe(4-NO_2) and Lys-Phe(4-NO_2) bonds were determined from the initial rate measurements, and the results are listed in Table 4. The mutations led to a 4- to 130-fold decrease of k_{cat}/K_m compared to the wild-type enzyme for peptide A. However, the k_{cat}/K_m value for the double mutant was larger than those of single mutants. Although the k_{cat}/K_m values of the double mutant for peptide B and C were 50 and 30 times smaller than for peptide A, respectively, the double mutation led to hydrolysis of the Lys-Phe(4-NO_2)

bonds in the former two peptides. The double mutant showed a similar ratio of k_{cat}/K_m for peptide C over peptide A to aspergillopepsin I, although the preference for peptide B was significantly different between the double mutated pepsin and aspergillopepsin I. These results indicated that fungal aspartic proteinase-like specificity of the T77D/G78(S)S79 mutant can be introduced into porcine pepsin (Table 4).

Table 4. Kinetic parameters for the hydrolysis of Phe-Phe(4-NO_2) and Lys-Phe(4-NO_2) bonds in the peptide substrates by wild-type and mutant pepsins

Each reaction was carried out in 0.1 M sodium acetate, pH 4.5, at 37°C

Mutant pepsin	K_m (mM)	k_{cat} (s^{-1})	k_{cat}/K_m ($mM^{-1} \cdot s^{-1}$)	Ratio[a]
PTEF↓F(4-NO_2)RL				
Wild type	0.04	61	1500	1.0
T77D	0.20	20	100	1.0
G78(S)S79	0.17	2.0	12	1.0
T77D/ G78(S)S79	0.06	23	380	1.0
Aspergillopepsin I	0.01	18	1800	1.0
PTEK↓F(4-NO_2)RL				
Wild type			-[b]	
T77D	0.18	0.10	0.56	0.0056
G78(S)S79		ND[c]		
T77D/G78(S)S79	0.15	1.1	7.3	0.019
Aspergillopepsin I	0.02	13	650	0.36
Ac-AAK↓F(4-NO_2)AA-NH_2				
Wild type		-[d]		
T77D		-[d]		
G78(S)S79			ND[c]	
T77D/G78(S)S79	0.16	2.1	13	0.034
Aspergillopepsin I	0.10	8.0	80	0.044

[a] Ratio is k_{cat}/K_m value relative to that with Pro-Thr-Glu-Phe-Phe(4-NO_2)-Arg-Leu as substrate in each enzyme.

[b] Cleavage occurred between Glu-Lys bond.

[c] No detectable cleavage was observed.

[d] Cleavage occurred between Phe(4-NO_2)-Ala bond.

Arrows (↓) indicate the scissible bonds.

(Shintani,T. *et al., J. Biol. Chem., 272,* 18855-18861 (1997))

To obtain information on recognition of a Lys residue at the P_1 position by the T77D/G78(S)S79 double mutant, the kinetic parameters for this double mutant were determined at various pH values using three peptide subsrates containing a Phe or Lys in P_1 (Figure 7). In all substrates, the K_m values were pH independent, and above pH 4 the values of k_{cat} and k_{cat}/K_m were controlled by the dissociation of carboxyl group with a pK_a of about 5. Below pH 4, however, the k_{cat} and k_{cat}/K_m values for the P_1 Lys substrates (peptides B and C) decreased as pH declined, while there was no significant change in k_{cat} and k_{cat}/K_m values for the P_1 Phe substrate (peptide A).

The pH-dependent changes of substrate preference are shown in Figure 8. The P_1 Lys/Phe preference increased as pH rose, which indicates that the dissociation of a carboxyl group as Asp77 may affect the preference of a Lys residue at the P_1 position.

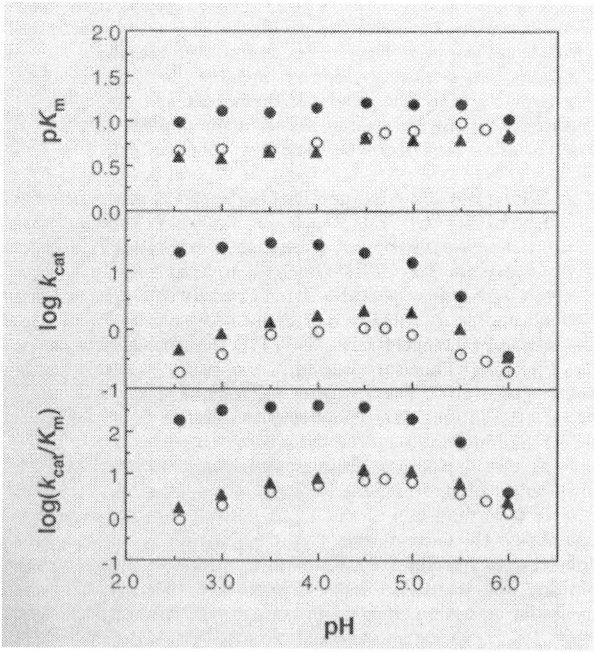

Figure7. pH dependence of pK_m. logk_{cat}, amd log(k_{cat}/K_m) for hydrolysis of Pro-Thr-Glu-Phe-Phe(NO_2)-Arg-Leu (●), Pro-Thr-Glu-Lys-Phe(NO_2)-Arg-Leu (○), and Ac-Ala-Ala-Lys-Phe(NO_2)-Ala-Ala-NH_2 (▲) by T77D/G78(S)S79 mutant pepsins.

(Shintani, T. *et al., J. Biol. Chem., 272*, 18855-18861 (1997))

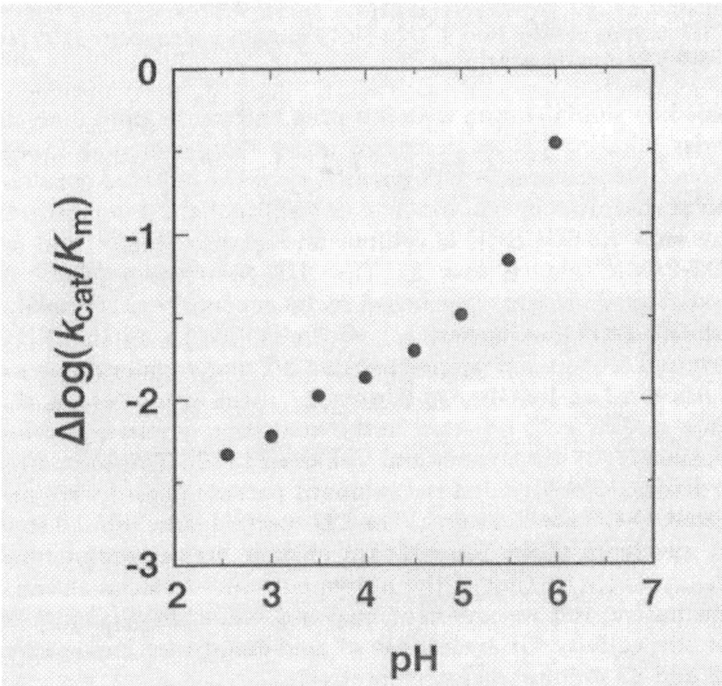

Figure 8. pH dependent substrate preference of double mutated pepsin. The P_1 Lys/Phe preference is represented by △log(k_{cat}/K_m). △log(k_{cat}/K_m) – log(k_{cat}/K_m)peptide B – log(k_{cat}/K_m)peptide A.

(Shintani, *et al., J. Biol. Chem., 272*, 18855- 18861 (1997))

In this study, the insertion of Ser between Gly78 and Ser79 of porcine pepsin may have caused an alteration in the direction of the Thr77 side chain and disabled the side chain of Thr77 from hydrogen bonding to the main chain of the substrate (Figure 5). These would result in overall decreases in catalytic efficiencies for all substrates in the G78(S)S79 mutant. In the T77D mutant, the presence of the longer Asp77 side chain may allow the flap to bind more loosely to the substrate and result in a decrease in the number of van der Waals contacts between enzyme and substrate; this would lead to a reduction in catalytic efficiency. However, the T77D mutant enzyme hydrolyzed the Lys-Phe(4-NO_2) bond in Pro-Thr-Glu-Phe(4-NO_2)-Arg-Leu (peptide B) although its catalytic efficiency was very low, whereas the Glu-Lys bond was the main cleavage site in wild-type pepsin. This would be due to the electrostatic repulsion between Glu of the substrate and Asp77 of the T77D mutant enzyme, and it would not be positive recognition of anionic Lys residue in P_1.

The wild-type and T77D mutant enzymes cleaved the Phe(4-NO_2)-Ala bond in Ac-Ala-Ala-Lys-Phe(4-NO_2)-Ala-Ala-NH_2 (peptide C) due to the strong interaction between the Phe(4-NO_2) residue of the substrate and S_1 hydrophobic pocket of the mutants. The double mutated pepsin exhibited higher catalytic efficiency (k_{cat}/K_m) for the P_1 Phe substrate than did the single mutants. The double mutant would have a flap structure and hydrogen bond network between Asp77 and the main chain of a substrate similar to those of the fungal aspartic proteinases. In the double mutant, the scissible peptide bond of the substrate may lie at a more stable position. Furthermore, the double mutant effectively hydrolyzed the Lys-Phe(4-NO_2) bond in peptide B and C, due to the additional two hydrogen bonds; the side chain of Asp77 and inserted Ser in the mutant may be hydrogen-bonded to the ε-amino group of the Lys residue in P_1 as fungal enzymes. In a previous study on aspergillopepsin I (30), the effect of the substitution of Ser79 by Ala (S79A) and the deletion of Ser79 (ΔS79; corresponding to porcine pepsin) on the substrate specificity were analyzed. The P_1 Lys/Phe preference of the ΔS79 mutant was reduced more significantly than the S79A enzyme. Therefore, in mutant pepsins the insertion of Ser is important for the orientation of the side chain at position 77, although this Ser may also contribute to the hydrogen acceptor in the hydrogen-bonding interaction.

The effect of the dissociation of the β-carboxyl group of Asp77 in double mutated pepsin on the recognition of Lys residue at the P_1 position was examined by pH-dependent hydrolysis of two types of peptide substrates containing Phe or Lys at P_1. The major differences in the pH activity profiles were the k_{cat}/K_m values below pH 4. These values for P_1 substrates (peptide B and C) decreased as the pH dropped, while there was no significant change for P_1 substrate (peptide A). The K_m values were pH independent in all cases. These results are different from those of the studies in which the existence of Lys or Arg residues at the P_4, P_3, P_2, P_3', P_4', and P_5' positions in the substrates influence the K_m values with little effect on k_{cat} in porcine pepsin [41]; the ionic interactions between the basic residues in the substrate and the side-chain carboxylates in the S_4, S_3, S_2, S_3', S_4', and S_5' subsites contribute to substrate affinity.

In double mutated pepsin, it is likely that the dissociation of Asp77 side chain contributes to the stabilization of the transition state when substrates contain Lys residue at the P_1 position rather than substrate affinity. The electrostatic interaction between the Asp77 side chain and P_1 Lys of the substrates could play an important role in recognition of Lys residue at the P_1 position. In aspergillopepsin I, however, the substitution of Asp77 by Asn hardly affected the P_1 Lys/Phe preference compared with wild-type enzyme, whereas the substitution of Asp77 by other amino acids (Glu, Ser, and Thr) and the deletion of Ser79 reduced the P_1 Lys/Phe preference with the decrease in the k_{cat} values for the P_1 Lys substrates. This indicates that the recognition of the basic P_1 side chain is not dependent on the specific electrostatic interactions between the P_1 side chain and Asp77 side chain of the enzyme [30]. From a site-directed mutagenesis study of rhizopuspepsin, Lowther *et al.* [31] concluded that the presence of Asp at position 77 has the potential to establish an extensive hydrogen-bonding network between the enzyme and the substrate containing Lys residue at the P_1 position. In double mutated pepsin, although the dissociation of the β-carboxyl group of Asp77 would be dispensable to accommodate the Lys residue in the S_1 pocket, it enhances the stabilization of the transition state complex when substrates contain a Lys residue at the P_1 position. This may be due to the formation of the electrostatic interaction or the changes of the hydrogen-bonding pattern between Lys residue in P_1 and Asp77 of the mutant pepsin as the pH is higher.

In conclusion, the double mutant pepsin T77D/G78(S)S79 was also able to activate bovine trypsinogen to trypsin by the selective cleavage of the Lys6-Ile7 bond of trypsinogen. The results of this study suggest that the structure of the active site flap contribute to the S_1 substrate specificity for basic amino acid residues in aspartic proteinases.

Chapter 4

AORSIN FROM *ASPERGILLUS ORYZAE*, A NOVEL SERINE PROTEINASE WITH TRYPSINOGEN ACTIVATING SPECIFICITY AT ACIDIC PH

4.1. SEDOLISIN

In addition to the above described aspartic proteinases, there exist some other distinct families that do not contain the consensus sequence of Asp-Thr/Ser-Gly. These are called non-pepsin-type aspartic proteinases and include aspergillopepsin II (EC 3.3.23.19) [42], scytalidiopepsin B (EC 3.4.23.32) [43], pseudomonapepsin (EC 3.4.23.37) [44] and xanthomonapepsin (EC 3.4.23.33) [45, 46]. These families of aspartic proteinases are not sensitive to pepstatin, hence they are also called pepstatin-insensitive carboxyl proteinases.

However, another important member of the aspartic proteinase (of family A7) is the classical late-infantile neuronal ceroid lipofuscinosis (LINCL) protein, CLN2p (ceroid lipofuscinosis, neuronal 2 protein), a human lysosomal enzyme that, when mutated, leads to the fatal neurodegenerative disease, LINCL [47-49]. CLN2p has both tripeptidyl-peptidase I and endopeptidase activity at acidic pH [50, 51]. The sequence of CLN2p is significantly similar to the two bacterial pepstatin-insensitive carboxyl proteinases, pseudomonapepsin and xanthomonapepsin [43-46]. Within the sequences of these enzymes, Rawlings and Barrett [50] identified a significant conserved sequence, Gly-Xaa-Ser, which is characteristic of the active-site motif of many serine peptidases. More recently, Lin *et al.* [52] have shown that Ser475,

Asp360 and Asp517 are essential for the activity of CLN2p/tripeptidyl-peptidase I, and also that diisopropyl fluorophosphates (DFP) specifically react with CLN2p at Ser475. Furthermore, the crystal structure of psedomonapepsin has been determined [53]. On the basis of its three-dimensional structure, Wlodawer *et al.* [53] have proposed that a family containing psedomonapepsin can be defined as a novel family of serine-like enzymes that have a unique catalytic triad, Glu80, Asp84 and Ser287, and the enzymes in this family are called serine-carboxyl proteinase [pseudomonapepsin is thus also known as *Psedomonas* serine-carboxyl proteinase (PSCP)]. This new family, 'sedolisin', was classified as S53 in the MEROPS database [54]. Consequently, the EC numbers for pseudomonapepsin and xanthomonapepsin have been reassigned as EC 3.4.21.100 and EC 3.4.21.101, respectively.

4.2. Aorsin from *A. oryzae*

An enzyme that hydrolyses Z-Arg-Arg-MCA at acidic pH was found in commercial protease M powder that was drained from a solid culture filtrate of *A. oryza*e [55]. We called the enzyme aorsin. The specific activity of the purified aorsin was 12.8 mkat/kg of protein for Z-Arg-Arg-MCA at 30°C and pH 4.0. Aorsin also hydrolysed clupeine and salmine as well as some fluorogenic peptide substrates, Boc-Gln-Arg-Arg-MCA and Boc-Leu-Lys-Arg-MCA, at an optimum pH of 4.0. Some enzymic and physicochemical properties are summarized in Table 5. The molecular mass of the enzyme was estimated to be 90 kDa and 70-120 kDa by gel filtration and SDS-PAGE (10% gel), previously. Protein that was deglycosylated by *Flabobacterium* N-glycosidase F migrated as a sharp single 61 kDa band using SDS-PAGE (10% gel). The results indicated that the enzyme is a monomer with hetero-N-gycosidases. The enzyme lost activity irreversibly at pH values of > 6.5 and < 3 during incubation at 30°C. The half-life ($t_{1/2}$) values for activity loss at 30°C were 150 min at pH 6.6, 5.5 min at pH 6.8 and 14 min at pH 2.4. Although the enzyme that was treated at 50°C and pH 4.0 for 10 min remained fully active, the residual activity of the enzyme treated at 55°C for 10 min was 65%. The $t_{1/2}$ value for loss of activity at 60°C and pH 4.0 was 3.5 min. The isoelectric point was judged to be 3-4 due to diffuse migration. Aorsin contained 1 mol of calcium per mol. The T_m was 66.7°C. To investigate the inhibitor of aorsin, the enzyme was treated with various inhibitors at 30°C and pH 4.0 for 30 min and the remaining activity was measured. The enzyme activity was inhibited neither by pepstatin nor the standard inhibitors of peptidases of various type, such as diazoacetyl-DL-norleucine methyl ester (DAN) (0.0% nM), E-64 (0.1 mM), iodoacetamide (1 mM), PMSF (10 mM), pefabloc SC (10 mM), EDTA (5 mM), EGTA (5 mM) or *o*-phenathroline (2 mM), at the final concentration indicated. This inhibition pattern resembled that of psedomonapepsin (or PSCP), a pepstatin-insensitive carboxyl-proteinase. Only antipain and leupeptin completely inhibited the aorsin at 0.01 mM, but aorsin was not inhibited by tyrostatin (56), which is a specific inhibitor of PSCP.

Table 5. Some enzymic and physicochemical properties of aorsin

Property	Method, symbol of condition	Value
Molecular mass	Gel filtration	90 kDa
	SDS-PAGE	7—120 kDa
	SDS-PAGE (deglycosylated)	61 kDa
	Deduced from nucleotide sequence	46,522 Da
Isoelectric point	p*I*	3-4
Ca^{2+} content	Atomic absorption spectrophotometer	1 mol/mol
Melting temperature (T_m)	Differntial scanning calorimeter	66.7°C
Optimum pH	pH	4.0
Specific activity	Z-Arg-Arg-MCA, 30°C, pH 4.0	12.8 mkat/kg
Half-life ($t_{1/2}$) for tolelance of pH	30°C, pH 6.6	150 min
	30°C, pH 6.8	5.5 min
	30°C, pH 2.4	14 min
Half-life ($t_{1/2}$) for tolerance of temperature	60°C, pH 4.0	
Inhibitor	Antipain, leupeptin	3.5 min

(Lee, B. R. *et al., Biochem. J., 371*, 541-548 (2003))

To investigate the substrate specificity of aorsin, 45 types of fluorogenic peptide substrates were used. Aorsin hydrolysed the substrates containing the basic amino acid residues arginine or lysine at the P_1 position, but did not hydrolyse any other substrates possessing non-basic amino acids at the P_1 position, indicating that the enzyme has trypsin-like activity. Several good fluorogenic substrates and their kinetic parameters are shown in Table 6. Although no significant difference in k_{cat}/K_m values was observed between the substrates, relatively low k_{cat}/K_m values and high K_m values were observed for Boc-Gln-Gly-Arg-MCA and Boc-Leu-Gly-Arg-MCA in which the P_2 positions are occupied by glycine. Low K_m values were also observed for substrates having dibasic amino acids at the P_1 and P_2 positions, such as Boc-Gln-Arg-Arg-MCA, Boc-Leu-Arg-Arg-MCA and Boc-Gly-Arg-Arg-MCA. The highest k_{cat}/K_m value was observed for Boc-Leu-Lys-Arg-MCA. On the other hand, aorsin hydrolysed a peptide hormone dynorphin A (Tyr-Gly-Gly-Phe-Leu-Arg-Arg-Ile-Arg-Pro-Lys-Leu-Lys) at the peptide bond between Arg7 and Ile8, but other peptide bonds were not cleaved despite the presence of several basic residues. These findings indicate that aorsin has a unique, specific requirement for basic residue at P_1 and to prefer paired basic residues. In terms of cleavage, aorsin is closer to a processing enzyme than it is to trypsin.

Table 6. Kinetic parameters of aorsin towards fluorogenic peptide substrates at pH 4.0.

The parameters were determined as described in the Experimental section of references [55].The k_{cat} value was calculated using molecular mass 46,522 Da. The values of k_{ca}t, K_m and k_{cat}/K_m are the means ± S.D. for three to five independent experiments.

Substrate	K_m (μM)	k_{cat} (s^{-1})	k_{cat}/K_m ($M^{-1} \cdot s^{-1}$)
Boc-Gln-Arg-Arg-MCA	0.7 ± 0.03	0.10 ± 0.01	(1.45 ± 0.06) x 10^5
Boc-Leu-Arg-Arg-MCA	3.9 ± 0.25	0.24 ± 0.02	(6.12 ± 0.45) x 10^4
Boc-Gly-Arg-Arg-MCA	2.8 ± 0.22	0.25 ± 0.02	(9.01 ± 0.47) x 10^4
Boc-Leu-Lys-Arg-MCA	10.0 ± 0.51	5.80 ± 0.37	(5.79 ± 0.38) x 10^5
Boc-Gly-Lys-Arg-MCA	26.8 ± 1.33	2.69 ± 0.18	(1.0 ± 0.07) x 10^5
Boc-Leu-Tyr-Arg-MCA	58.0 ± 2.47	4.11 ± 0.12	(7.09 ± 0.38) x 10^4
Boc-Leu-Gly-Arg-MCA	84.5 ± 3.75	1.26 ± 0.04	(1.49 ± 0.09) x 10^4
Boc-Gln-Gly-Arg-MCA	88.3 ± 4.06	0.84 ± 0.05	(9.46 ± 0.07) x 10^3

(Lee, B. R. *et al., Biochem. J., 371*, 541-548 (2003))

Plasminogen that was treated with aorsin hydrolysed a Boc-Val-Leu-Lys-MCA substrate, substrate of plasmin, whereas aorsin or plasminogen alone did not hydrolyse the substrate. However, we failed to determine the processing site because the activated plasmin could not be purified, probably due to further degradation. Trypsinogen was also activated by being treated with aorsin, and the generated trypsin hydrolysed Z-Phe-Arg-MCA, which was not cleaved by aorsin. The NH_2-terminal sequence of the activated trypsin was determined to be Ile-Val-Gly-Gly-Tyr-, indiating that accurate processing had occurred by aorsin as shown in Figure 9.

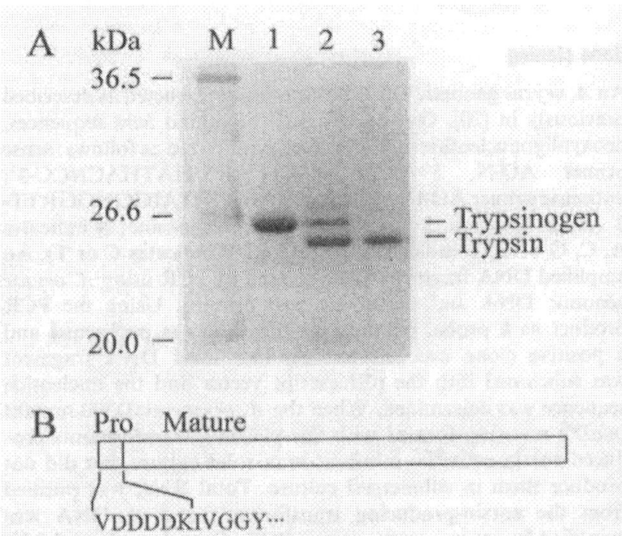

Figure 9. Conversion of trypsinogen into trypsin by aorsin. (A) The reaction mixtures for the trypsinogen activation were analyzed by SDS-PAGE on a 15% gel and stained with Coomassie Brilliant Blue R-250. Trypsinogen was stained with apparent homogeneity (*lane 1*). The reaction product, convert into trypsin, migrated slightly faster than trypsinogen (*lane 2*), and this reaction was similar to that of authentic trypsin (*lane 3*). (B) The natural processing site of bovine trypsinogen (24 kDa). The N-terminal sequence of the converted trypsin (lane 2 in A) was determined to be IVGGY.

(Lee, B. R. *et al., Biochem. J., 371*, 541-548 (2003))

A gene (*aorO*) was cloned from *A. oryzae* genomic library. To confirm whether the gene is functional, a plasmid containing a 4 kb *Xba*I fragment was used to transform the *A. oryzae* niaD300(*niaD*$^-$) mutant strain. Consequently the enzyme activity of the transformant was observed to be about 10-fold higher than that of the recipient strain in the solid culture, but neither demonstrated any activity in the submerged culture. This finding indicates that the cloned gene is functional and expressed only in the solid culture state.

A 4 kb *Xba*I fragment of the cloned *aorO* gene and cDNA were sequenced and compared. Eight introns with short sequences of 50-75 nucleotides were found in the gene. The complete structural gene encoded a protein of 652 amino acids. The amino acid sequences of the lysyl-endopeptidase-digested peptides were verified at positions Asp516 to Leu530 and Asn604 to Thr612. The NH_2-terminal amino acid sequence of the mature enzyme was found at positions Gly216 to Pro244. Consequently, the pre-propeptide consisted of 215 amino acid residues from Met1 to Lys215, and the mature form of the enzyme was composed of 437 amino acid residues. The deduced amino acid sequence of the mature protein contained six potential sites for aspargine-linked glycosylation at positions 218, 247, 331, 445, 604 and 613. A protein identify search revealed that the mature aorsin protein was 35% identical with CLN2p [47], 23% identical with PSCP [44] and 22% identical with *Xanthomonas* pepstatin-insensitive carboxyl proteinase (XCP) [46], all of which are pepstatin-insensitive carboxyl proteinases. Several significant amino acids were conserved in aorsin at Glu86, Asp90, Asp211, Ser354 and Asp395, each of which was equivalent to Glu80, Asp84, Asp170, Ser287 and Asp328 in PSCP, respectively, and all of which have previously been described as residues essential for enzyme activity [53, 57].

To identify the catalytic residues, the *aorO* gene was mutated at the site of encoding Glu86, Asp90, Asp211, Ser354 or Asp395. Since we confirmed that an authentic *aorO* gene product, aorsin, was not detected in the submerged culture broth of *A. oryzae*, we employed *A. oryzae* as a host for recombinant aorsin production. Purified recombinant enzymes showed that they have the same patterns as authentic aorsin in terms of molecular mass, carbohydrate moiety and NH_2-terminal sequence. The results of site-directed mutagenesis of aorsin are shown in Table 7.

An aspartic acid residue, Asp328 in PSCP and Asp517 in CLN2p, has been identified as catalytic residue [52, 57] or a ligand of calcium binding [53]. However, mutant aorsins, Asp395→Asn (D395N) and Asp395→Gln (D395Q), had full activity and lost their calcium-binding ability completely. This finding indicates that Asp395 is a calcium-binding site and is not related to the catalytic function. A mutant enzyme replacing Asp211, equivalent to Asp170 in PSCP and Asp360 in CLN2p, with either asparagines (Asp211→ Asn, D211N) or glutamic acid (Asp211→Glu, D211E) showed that the k_{cat}/K_m

values were remarkably reduced by 7800-42000-fold. Mutation of Ser354, which is equivalent to Ser287 in PSCP and Ser475 in CLN2p, with threonine (Ser354→Thr, S354T) reduced the k_{cat}/K_m value by 17000-fold. Substitution of Glu86, equivalent to Glu80 in PSCP, with either aspartic acid (Glu86→Asp, E86D) or glutamine (Glu86→Gln, E86Q) also reduced the k_{cat}/K_m value by 1000-7400-fold. Surprisingly, however, the k_{cat}/K_m value of the replacement of Asp90, equivalent to Asp84 in PSCP, with asparagine (Asp90→Asn, D90N) was only 21-fold lower than the k_{cat}/K_m value of the wild-type enzyme.

Table 7. Kinetic parameters for hydrolysis of fluorogenic peptide substrate Boc-Leu-Lys-Arg-MCA catalysed by aorsin and its mutants at pH 4.0

Mutant aorsin	K_m (μM)	k_{cat} (s^{-1})	k_{cat}/K_m ($M^{-1} \cdot s^{-1}$)	Relative k_{cat}/K_m
Wild type	10.0 ± 0.51	(5.80 ± 0.37)	$(5.79 \pm 0.38) \times 10^5$	1
E86D	63.6 ± 4.80	$(3.74 \pm 0.22) \times 10^{-2}$	$(5.89 \pm 0.27) \times 10^2$	1×10^{-3}
E86Q	19.7 ± 0.23	$(1.55 \pm 0.02) \times 10^{-3}$	$(7.87 \pm 0.34) \times 10$	1.4×10^{-4}
D90E	117.0 ± 5.85	$(3.52 \pm 0.14) \times 10^{-2}$	$(3.00 \pm 0.17) \times 10^2$	5.2×10^{-4}
D90N	22.5 ± 1.87	$(6.15 \pm 0.31) \times 10^{-1}$	$(2.75 \pm 0.19) \times 10^4$	0.048
D211E	23.6 ± 1.51	$(3.23 \pm 0.21) \times$ 10-^{4}	$(1.37 \pm 0.05) \times 10$	2.4×10^{-5}
D211N	16.1 ± 1.38	$(1.19 \pm 0.05) \times 10^{-3}$	$(7.41 \pm 0.42) \times 10$	1.3×10^{-4}
S354T	19.3 ± 1.23	$(6.45 \pm 0.11) \times 10^{-4}$	$(3.34 \pm 0.25) \times 10$	5.8×10^{-5}
D395E	10.6 ± 0.92	(5.26 ± 0.26)	$(4.97 \pm 0.33) \times 10^5$	0.86
D395N	9.6 ± 0.62	(4.49 ± 0.09)	$(4.68 \pm 0.26) \times 10^5$	0.81

(Lee, B. R. *et al., Biochem. J., 371*, 541-548 (2003))

Consequently, the k_{cat}/K_m values of the mutant enzymes Glu86→Gln (E86Q), Asp211→Asn (D211N), and Ser354→Thr (S354T) were 3-4 orders of magnitude lower, and Asp90→Asn (D90N) was 21-fold lower than that of wild-type aorsin, indicating that the positions are important for catalysis. Aorsin is another of the S53 family serine-carboxyl proteinases that are not inhibited by pepstatin.

Chapter 5

DEUTEROLYSIN FROM *ASPERGILLUS*, A MEMBER OF ASPZINCIN FAMILY WITH A NEW ZINC-BINDING MOTIF (HEXXH + D)

5.1. DEUTEROLYSIN – EXTREAMLY HEAT STABLE 19 KDA ZN^{2+}-PROTEASE –

Deuterolysin (EC 3.4.24.39) is extremely stable at 100°C, but unstable around 75°C [58-61]. Deuterolysin shows unique thermal stability. It is most unstable after 10 min at about 65-75°C, but it regains stability beyond this temperature, and it is relatively stable at 100°C. This phenomenon is attributed to unique properties of deuterolysin in regard to reversible thermal unfolding and resistance for autoproteolysis [61].

Metalloendopeptidases are physiologically important proteinases for processing proteins in eukaryotes and prokaryotes. Hooper attempted to present a scheme based on the zinc-binding site, and this has been extended to classify zinc metalloproteinases into distinct families [62]. There are gluzincins, metazincins, inverzincins, the carboxypeptidase family, and the DD-carboxypeptidase family. Rawlings and Barrett divided the families of metallopeptidases into five groups based on the zinc-binding motif: "HEXXH + E", "HEXX + H", "HEXXH + ?", heterogenous other than HEXXH, and unknown [63]. In the clan MX, the identity of the third ligand to the zinc is unknown.

Site-directed mutagenesis on deuterolysin (EC 3.4.24.39, neutral proteinase II) from *A. oryzae* confirmed that His128 and His132 are zinc lignads and that Glu129 is a catalytically crucial residue [64]. These studies also identified aspartic acid residue (Asp164) as the third zinc ligand and, as a consequence, a new family name for deuterolysin containing the zinc-binding motif HEXXH + D was proposed: the 'aspzincin'.

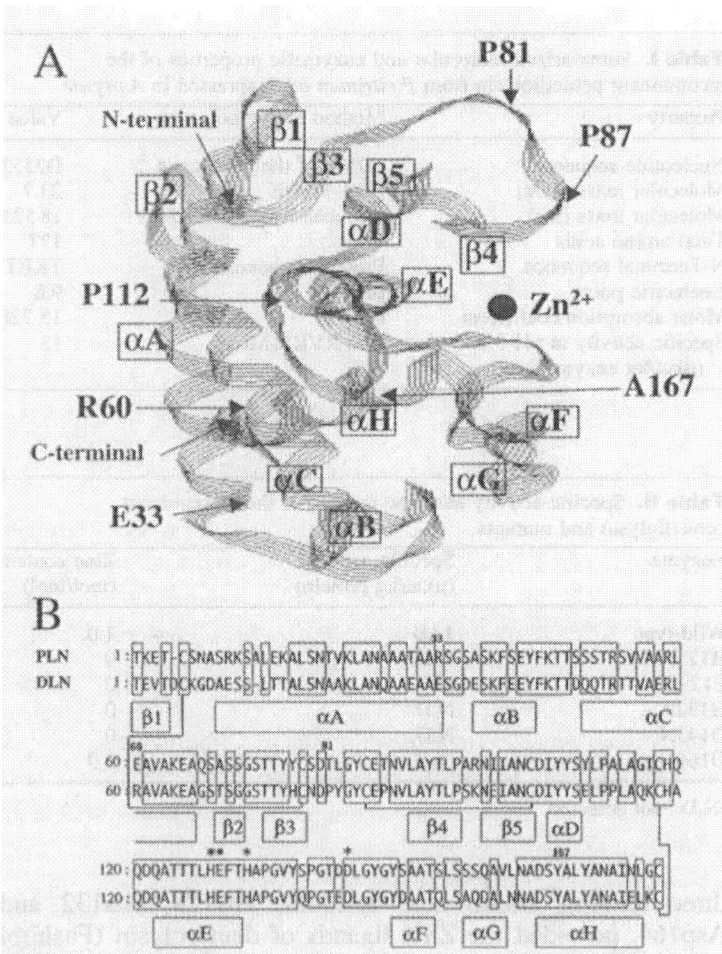

Figure 10. Secondary structure of deuterolysin reported by McAuley *et al*. (2001) (A) and sequence alignment of penicillolysin (PLN) and deuterolysin (DLN) (B).

(Doi. Y. *et al., Protein Eng. Des. Select., 17*, 261-266 (2004))

For the crystallographic analysis of McAuley and coworkers [65], deuterolysin is an αβ protein with a zincin-like fold. It comprises eight α-helices and four strands of β-sheet (Figure 10). There are three disulfide bridges between residues Cys6-Cys78, Cys85-Cys103, and Cys117-Cys177; the last of these involves the C-terminal residue, which bridges to a cysteine residue located on a loop region between helices *D* and *E*. The secondary structure of deutrolysin is shown in the text [65].

The structure of another aspzincin with 21% sequence homology to deuterolysin, a peptidyl-Lys-metalloendopeptidase from the fungus *Grifola frondosa*, has been determined [66]. This enzyme is closely related in structure to deuterolysin, with an r.m.s. difference in C^{α} positions of 1.2 Å and a topological diversity score of 8.4, yet it is still a poor molecular replacement search model.

The active site of deuterolysin is located in a large cleft on the protein surface formed by β-sheets and an α-helix. The histidine residues coordinating the zinc are located on this helix (α*E*). The Zn^{2+} atom is liganded by two histidine residues (His128 and His132), an aspartic acid residue (Asp143) and two water molecules, folding a distorted octahedral geography as shown in the reference [66]. The metal-ligand distances range from 2.05 and 2.35 Å, falling into the expected range for zinc metal geometry. Several of the angles within the octahedron deviate by more than 10° from ideal value and this distortion is a consequence of accommodation of the bidentate carboxylate group of Asp143 [65]. The catalytic residue which promotes the nucleophilic attack of a water molecule on the carbonyl C atom of the substrate is Glu129.

From the indication of McAurey [65], the structure confirms that deuterolysin is an aspzincin as proposed by Fushimi *et al.* [64], but there is a disagreement over the identitity of the aspartic acid that ligands the zinc. Fushimi and coworkers deduced this was Asp164 after their studies showed mutating this to an asparagines residue led to loss of the zinc ion, whereas the Asp143→Asn mutant retained the zinc. It is not clear why the Asp143 mutant retained the zinc; however, a suggestion as to why mutating Asp146 led to loss of the zinc can be made. Asp146 is located close to His128, with a distance of 2.86 Å between the carboxylate oxygen OD1 of Asp164 and nitrogen ND1 of the histidine. The loss of this interaction after mutation to asparagines leads to a change in the orientation or position of this histidine, resulting in loss of zinc-binding ability [65].

5.2. Specificity of Deuterolysin

Deuterolysin shows high activity on basic nuclear proteins such as histone and protamine but has low activities on substrates routinely used in the laboratory, such as casein, hemoglobin, albumin and gelatin [58, 59]. The action of deuterolysin on calf thymus histone H4 was studied [67]. Deuterolysin hydrolyzed histone with various amino acid residues at the P_1 position including basic arginine and lysine, acidic aspartic and glutamic acids, and proline. Deuterolysin had high activity towards peptides next to a pair of basic residues (Lys-Arg, Arg-His, Arg-Lys, and Arg-Arg) in the calf thymus histone H4 (Figure 11). Deuterolysin hydrolyzed bonds between Arg17-His18, His18-Arg19, Arg23-Asp24, and Arg36-Leu37. On the other hand, deuterolysin also hydrolyzed Pro32-Arg33, Glu63-Asn64, Asp68-Ala69, and Met84-Asp83. The specificity of deuterolysin was different from the specificity of microbial metalloendopeptidases such as thermolysin (EC 3.4.24.27) and *A. sojae* neutral proteinase I.

Deutrolysin had strong hydrolytic activity toward Boc-Arg-Val-Arg-Arg-MCA, with optimal activity at pH 9.0 (Table 8). The k_{cat}/K_m value of 833 s^{-1} M^{-1} was obtained for Boc-Arg-Val-Arg-Arg-MCA. The enzyme hydrolyzed other dibasic-MCA derivatives: Pyr-Arg-The-Lys-Arg-MCA, Boc-Leu-Lys-Arg-MCA, Boc-Leu-Arg-Arg-MCA, and Z-Arg-Arg—MCA. However, almost no hydrolysis by the enzyme was found toward mono or certain dibasic peptide-MCAs. The findings indicate that deuterolysin had substrate specificity toward paired dibasic amino acid residues at the P_2-P_1 positions.

Cleavage sites following pairs of basic residues:

GAKR^{17} ↓ H^{18}RKVLRDNIQGITKPA
AKRH^{18} ↓ R^{19}KVLRDNIQGITKPA
AIRR^{36} ↓ L^{37}ARRGGVK

Other cleavage sites:

KVLR^{23} ↓ D^{24}NIQGITKQ
ITKP^{32} ↓ A^{33}IRRLA
VFLE^{63} ↓ N^{64}VIRD
VIRD^{68} ↓ A^{69}VTYTEHAKRKTVTAMVV
VTAM^{84} ↓ D^{85}VVY

Figure 11. Summary of cleavage specificity of deuterolysin from *A. oryzae* toward calf thymus histone H14.

(Doi, Y. *et al., Biosci. Biotechnol. Biochem., 67*, 264-270 (2003))

The arrows (↓) indicate cleavage sites. The right hand sequences of the arrows indicate the N-terminal amino acid sequences of the small peptide obtained with an Applied Biosystems 743 protein sequencer.

5.3. Co-Deuterolysin

The value of electron paramagnetic resonance (EPR) spectroscopy in establishing structures and their dependence on metal coordination-sphere composition and geometry has been illustrated by its application to several aspects of cobalt bioinorganic chemistry.

Table 8. Kinetic parameters of deuterolysin and Co-deuterolysin toward fluorogenic MCA substrates

Substrate	Deuterolysin			Co-deuterolysin		
	K_m (μM)	k_{cat} (s^{-1})	k_{cat}/K_m ($s^{-1} \cdot M^{-1}$)	K_m (μM)	k_{cat} (s^{-1})	k_{cat}/K_m ($s^{-1} \cdot M^{-1}$)
Boc-RVRRMCA	26.4 ± 4.0	2.2 x 10^{-2}	833 (100*)	27.5 ± 10.0	1.7 x 10^{-2}	618 (74)
Pyr-RTKR-MCA	23.7 ± 7.4	2.2 x 10^{-3}	84 (10)	11.9 ± 1.8	8.2 x 10^{-4}	69 (8)
Boc-RVRR-MCA	146.0 ± 21.0	9.0 x 10^{-3}	62 (7)	157.3 ± 26.2	3.5 ± 10^{-3}	22 (3)
Boc-LRR-MCA	70.0 ± 3.5	4.5 x 10^{-3}	71 (9)	63.8 ± 5.7	1.7 x 10^{-3}	27 (3)
Z-RR-MCA	47.3 ± 10.0	9.0 x 10^{-4}	19 (2)	52.2 ± 5.7	6.5 x 10^{-4}	12 (1)

*The rate of hydrolysis of Boc-RVRR-MCA by deuterolysin is taken to be 100, and other such values are relative to that. Results are shown as means ± S.D. of three determinations.

MCA, 4-methylcoumacyl-7-amide.

(Doi, Y. *et al., Biosci. Biotechnol. Biochem., 67*, 264-270 (2003))

Deuterolysin from *A. oryzae* has a zinc-binding motif, but the zinc is not detected EPR. Removal of the zinc ion from *A. oryzae* deuterolysin yields the inactive apo-enzyme, which can also be reactivatd by stoichiometric amounts of zinc ions. Moreover, the addition of cobalt ions to the apo-enzyme results in an enzymatically active species. We purified Co-deuterolysin to examine cobalt coordination and geometry by EPR [67]. Kinetic parameters toward fluorogenic MCA substrates (Table 8) and the CD spectrum of cobalt-substituted deuterolysin (Co-deutrolysin) were similar to that of the native deuterolysin.

In Figure 12 we show the Q-band EPR spectra of Co-deuterolysin measured at 10 to 5 K. Preliminary observation of the X- and S-band EPR (9 and 3 GHz, respectively) showed an axial feature, but line-broadening prevented us from identifying the *g* principal values from the spectrum. Resolution improvement with respect to *g* values is one advantage of the use of a high microwave frequency. From the spectra at 33.9570 GHz, *g* principal values could be unambiguously measured.

The dominant line-broading mechanism of Co-deuterolysin was unresolved hyperfine splitting of cobalt nucleus and *g*-strain. The line width from a *g*-strain mechanism increases with increasing microwave frequency, but the contribution from hyperfine splitting is independent of the microwave frequency.

On the basis of the Q-band spectrum giving better resolution than that the X- and S-bands, we concluded that the dominant line-broadening mechanism at Q-band EPR was *g*-strain and that there was broadening in the X- and S-band spectra. From spectra simulation in the Q-band spectrum, we obtained the *g* principal values of $g_{xx} = 5.20$, $g_{yy} = 4.75$, and $g_{zz} = 2.24$, and the *g*-strain tensor principal values $dg_{xx} = 1.11$ (0.2), $dg_{yy} = 1.11$ (0.2), and $dg_{zz} = 0.19$ (0.01).

The absorption spectrum of Co-deuterolysin indicates that the number of coordinating ligands can be four or five, that is, tetrahedral or trigonal-bipyramid, respectively. As a four coordinate site, the absorption energy of Co-deuterolysin is a little high, and the absorption coefficient ε is small. On the other hand, the spectrum lacks the characteristic band (16,000 cm-1) of a five-coordinate site.

It can be concluded that a distorted tetrahedral or a two-triangular pyramidal environment has been assigned to the Co-deuterolysin from the correlation of the *g* principal values, $g_{xx} = 5.20$, $g_{yy} = 4.75$, and $g_{zz} = 2.24$ [67].

5.4. PENICILLOLYSIN – THERMOLABILE 19 KDA ZN^{2+}-PROTEASE –

Penicillolysin is a member of the clan MX and the family of M35 proteases [68-71]. Penicillolysin is a single-chain protein of 177 amino acid residues with a calculated molecular mass of 18525 Da and p*I* of 9.6, which contains three disulfide bonds (Cys6-Cys78, Cys85-Cys103, and Cys117-Cys177) (Figure 10). The enzyme is a thermolabile Zn^{2+}-protease from *Penicillium citrinum* with a unique substrate profile.

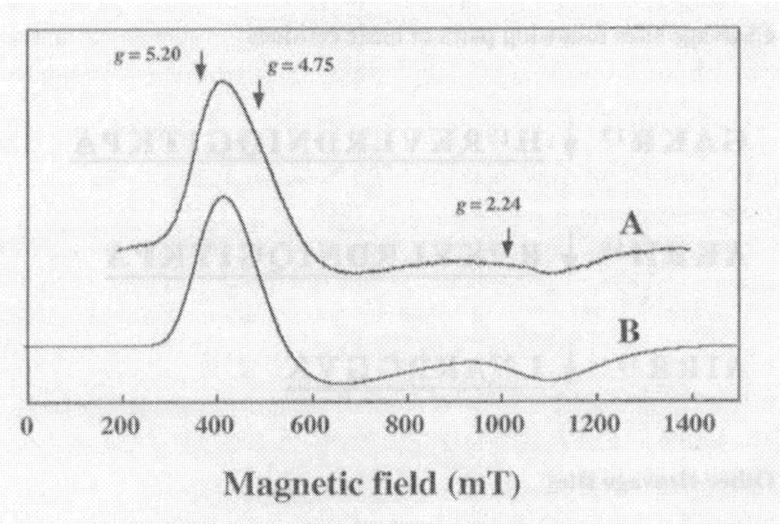

Figure 12. Q-band EPR spectra of Co-deuterolysin. A: spectrum observed. B: Simulated spectrum made with the *g* and *g*-strain tensor principal values.

(Doi, Y. *et al., Biosci. Biotechnol. Biochem., 67*, 264-270 (2003))

Penicillolysin showed a high affinity toward the Pro-X (X = Gln, Lys, Leu or Arg) bonds of substance P, dynorphin-A (1-13),neurotensin and chicken brain pentapeptide, and the R-R bonds in dynorphin A and neurotensin (Figure 13) [69]. Preferential cleavages of bonds by the enzyme with hydrophobic amino acid residues at the P_1 position were observed on the peptides used (Figure 13). The specificity of penicillolysin differs from that of other metalloprotinases.

The specific activities of penicillolysin for clupeine and casein hydrolysates were 3.04 x 10^{-1} and 5.23 x 10^{-3} katal/kg protein at pH 7.0, respectively (Table 9) [69]. The rate of clupeine hydrolysis was 60-fold greater than that for casein hydrolysis. When zinc is removed, the enzyme is completely inactive, and readdition of zinc restores the dual activities towards clupeine and casein (Table 9). Depending on the casein substrate, the cobalt-penicillolysin (Co-penicillolysin) could be up to ca 1.6 times more active than the native zinc enzyme. On the other hand, in clupeine-hydrolysis, the cobalt enzyme is about 70% as active as the native enzyme. Thus, replacement of the zinc-penicillolysin with cobalt markedly decreases the activity towards clupeins while increases it towards casein.

The curve in Figure 14 shows the CD spectrum of penicillolysin [69]. The α-heix and β-structure content of penicillolysin were ca 19 and 58%, respectively. The CD curve of the apoenzyme is in Figure 14B and is notably different from the native one. The α-helix and the β-structure content of the apoenzyme were ca 9 and 61%, respectively. About 50% of the α-helix was destroyed in the conformational changes from the native to apoenzyme. The CD curve of Co^{2+}-reconstructed enzyme, Co-penicillolysin, is shown in Figure 14C. The contents of α-helix and β-structure were ca 20% and 58%, respectively. This CD curve was almost the same as that the native enzyme. Parameters of the secondary structure of penicillolysin are shown in Table10.

Substance P	**RPKP ↓ QQFF ↓ G ↓ LM**
Dynorphin-A	**YGGFLR ↓ RIRP ↓ KLK**
Neurotensin	**<QL ↓ YENKP ↓ R ↓ RPYIL**
Chicken brain pentapeptide	**LP ↓ LRF$_{NH_2}$**
Bradykinin	**RPPGF ↓ SPFR**
LHRH	**<QHW ↓ SY ↓ GLRPG$_{NH_2}$**
α-Melanotropin	**AcSY ↓ SMEHF ↓ RW ↓ GKPV$_{NH_2}$**
α-Neoendorphin	**YGGF ↓ L ↓ RKYPK**
Angiotensin II	**DRVYIHPF**
Angiotensin I	**DRVYIHPFHL**
CCK-octapeptide (26-33)	**DYMGWMDF$_{NH_2}$**

The arrows in the sequences indicate cleavages for 1 h hydrolysis by the enzyme. The degree of hydrolysis being: ↓ > ↓ > ↓ .

Figure 13. Summary of cleavage specificity of penicillolysin towards various oligopeptides.

(Yamaguchi, M. *et al., Phytochemistry, 33*, 1317-1321 (1003))

Table 9. Specific activities of penicillolysin, and Zn^{2+}- and Co^{2+}-penicillolysin at pH 7.0

Enzyme	Specific activity (katal · kg^{-1} protein)	
	Clupeine	Casein
Native	3.04×10^{-1} (100)	5.23×10^{-3} (100)
Zn^{2+}-penicillolysin	2.85×10^{-1} (94)	5.20×10^{-3} (99)
Co^{2+}-penicillolysin	2.10×10^{-1} (69)	8.26×10^{-3} (158)

(Yamaguchi, M., *et al., Phytochemistry, 33*, 1317-1321 (1993))

Several candidates for zinc ligands of penicillolysin were mutated and confirmed [72]. Substitutions of His128 and His 132 with Arg, of Glu129 with Gln, of Asp143 with Asn and of Asp164 with Asn resulted in a complete loss of both the enzyme activity and zinc binding ability of all mutant enzymes except D164N. As for the mutant D164N, the zinc content was determined as 1.0 mol/mol of the enzyme and the specific activity was found to be 53%. This finding indicates that the Zn^{2+} atom is bound by two histidine residues (His128 and His 132) and an aspartic acid residue (Asp143). Substitution of Glu129, a catalytically crucial residue in penicillolysin, with glutamine resulted in a mutant enzyme (E129Q) which exhibited no activity.

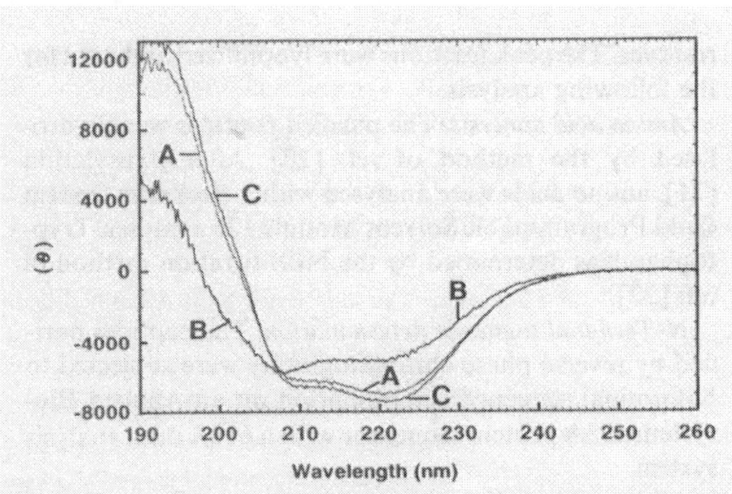

Figure 14. CD spectrea of Penicillolysin(A), apoenzyme (B), and Co2+-penicillolysin (C).

(Yamaguchi, M. *et al., Phytochemistry, 33*, 1317-1321 (1993))

Table 10. Secondary structure content (%) in penicillolysin, apoenzyme and Co^{2+}-penicillolysin

Enzyme	α-Helix	β-Structure	Turn	Random
Penicillolysin	19	58	3	21
Apoenzyme	9	61	2	28
Co^{2+}-penicillolysin	20	58	2	20

(Yamaguchi, M. *et al., Phytochemistry, 33*, 1317-1321 (1993))

A full-length cDNA encoding the penicillolysin, an 18 kDa metallo-endopeptidase from *Penicillium citrinum*, was cloned [70]. Analysis of the 1284 base pair nucleotide sequence of the cDNA revealed a single open reading frame coding for 351 amino acid residues. The coding region of penicillolysin gene, *plnC*, occupies 1053 base pairs of the cDNA. The sequence consists of a putative 19-residue signal sequence, a 155-residue propeptide segment, and the 177-residues of penicillolysin with a molecular mass of 18,529. Penicillolysin has a sequence of 68% with that of deuterolysin from *A. oryzae*.

Penicillolysin is thermolabile and its activity is almost completely lost above 66°C. It is interesting that, despite having a highly similar identity to and the same number, three, of disulfide bonds as deuterolysin, penicillolysin is thermolabile. Of the substitutions involving residues with largely identical structural environments in penicillolysin and deuterolysin, those for which the deuterolysin situation could be favorable for thermal stability were selected [72].

We designed and constructed two different mutants (R33E/E60R and A167E) for the connection between the two α-helices (A, and C helices or E, and H helices) in the folded penicillolysin [72]. The approximate average content of α-helices and β-structures in a protein can be estimated from the CD spectrum, by comparison to a database of spectra of proteins of known structure. Although the CD spectra of the wild-type and mutants have slight differences, the enzyme activity of the mutants was not largely affected. Kinetic parameters of the mutant R33E/E60R are similar to those of the wild-type enzyme. Similar k_{cat}/K_m parameters were obtained for the two mutants, R33E/E60R ($18.2\ M^{-1} \cdot s^{-1}$) and T81P ($16.8\ M^{-1} \cdot s^{-1}$). The k_{cat}/K_m value of the A167E mutant ($29.8\ M^{-1} \cdot s^{-1}$) was 1.6-fold higher than that of the wild-type enzyme ($18.2\ M^{-1} \cdot s^{-1}$).

The stability of both the wild-type and mutant proteins is expressed as the melting temperature, T_m, which is the temperature at which 50% of the enzyme is denatured during irreversible heat denaturation. To prevent the self-digestion of penicillolysin, heat treatment was carried out at pH 5.0 because the proteolytic activity is substantially reduced at this pH and the active form of the enzyme is stable. For the wild-type enzyme and the three mutants, the thermal stabilities and the T_m changes were assayed at pH 5.0 by measuring the far-UV CD spectrum at 222 nm as a function of temperature. For the wild-type penicillolysin, the T_m value was determined as 68°C at pH 5.0. The chosen mutation in penicillolysin was T81P, which caused an increase in the melting temperature to 71°C at pH 5.0. Unfortunately, we could not obtain thermal stable forms for the two mutants, T78P and A112P, of penicillolysin. The two single mutants, T78P and A112P, were found to be fairly unstable enzymes. The destabilizing effect of T87P is probably due to the location of Thr87, which is near the disulfide bridge, Cys85 and Cys103. Ala112 is located in a loop between αD and αE. The αE helix includes a catalytic residue, Glu129, and Zn^{2+}-binding residue, His128. When Ala112 is substituted with a more rigid residue, Pro, a minor structural change to the loop and the αE helix regions might occur. The chosen mutations in penicillolysin, R33E/E60R and A167E, caused an increase in the T_m to 70°C at pH 5.0.

When the rate of denaturation of enzyme is measured, it is more common to express the result in terms of the half-life of enzyme activity, $t_{1/2}$, rather than in terms of the rate constant, k_d. For a first-order reaction the relationship is $t_{1/2} = \ln 2/k_d = 0.69/k_d$ [73]. In our study, thermal stability was quantified as $t_{1/2}$, being the temperature at which 50% of the initial activity was retained for 10 min. The thermal stability of the mutants also depended on temperature, as shown in Figure 15. The double mutant R33E/E60R was found to be fairly thermally stable, as shown in Figure 15. The half-life ($t_{1/2}$) value, 66°C, of R33E/E60R was higher than that of the wild-type penicillolysin. It is interesting that two mutations, probably due to favorable interactions between the two α-helices (A and C helices), were generated and these interactions are important in the thermal stability of the mutant enzyme R33E/E60R. The A167E mutant showed the same thermal stability as R33E/E60R. The A167E mutation was also sufficient to increase the $t_{1/2}$ value from 62 to 65°C. The single A167E mutation was useful for the thermal stabilization of penicillolysin, indicating that the connection between the αE and αH helices may be important.

Among the three proline mutants, the T81P mutant significantly increased the $t_{1/2}$ value from 62 to 68°C, whereas T87P and A112P had no clear stabilizing effects. For the 50% residual activity, the thermal stability at 65°C was 4, 12, 12 and 21 min for the wild-type, R33E/E60R, A167E and T81P, respectively (Figure 15B). When the T81P mutation was introduced, the enzyme was found to be more thermally stable than the two mutants, R33E/E60R and A167E. It is interesting that T81P similarly enhanced the thermal stability of penicillolysin; however, the other two mutants, T87P and A112P, showed decreased thermal stability. The T81P mutant of penicillolysin may have a structure essentially identical to that of the wild-type of deuterolysin. There are indications that the proline residues contribute to enzyme thermal stability [74, 75]. The contribution can be explained in terms of entropic effects (or chain flexibility) since proline is more restricted conformationally in the unfolded state than any other amino acid.

Since proline residues are restricted with respect to their φ-ψ angles, their introduction can easily cause conformational strain in the backbone. Suzuki *et al.* [76] demonstrated that an increase in the frequency of proline in β-turns and in the total number of hydrophobic residues can enhance proline thermal stability.For further investigation, the triple mutant R33E/E60R/T81P was constructed. Surprisingly, the triple mutation was found to produce a fairly thermally unstable enzyme. The half-life ($t_{1/2}$) value of the triple mutant was determined as 58°C.

The thermal stability of an enzyme is influenced by many factors, including amino acid sequence, three-dimensional structure, cofactors and pH. However, at present we cannot explain the results obtained. The present findings described in the references [72] show that the thermal stability of penicillolysin is increased dramatically by just a few mutations in one particular region of the protein.

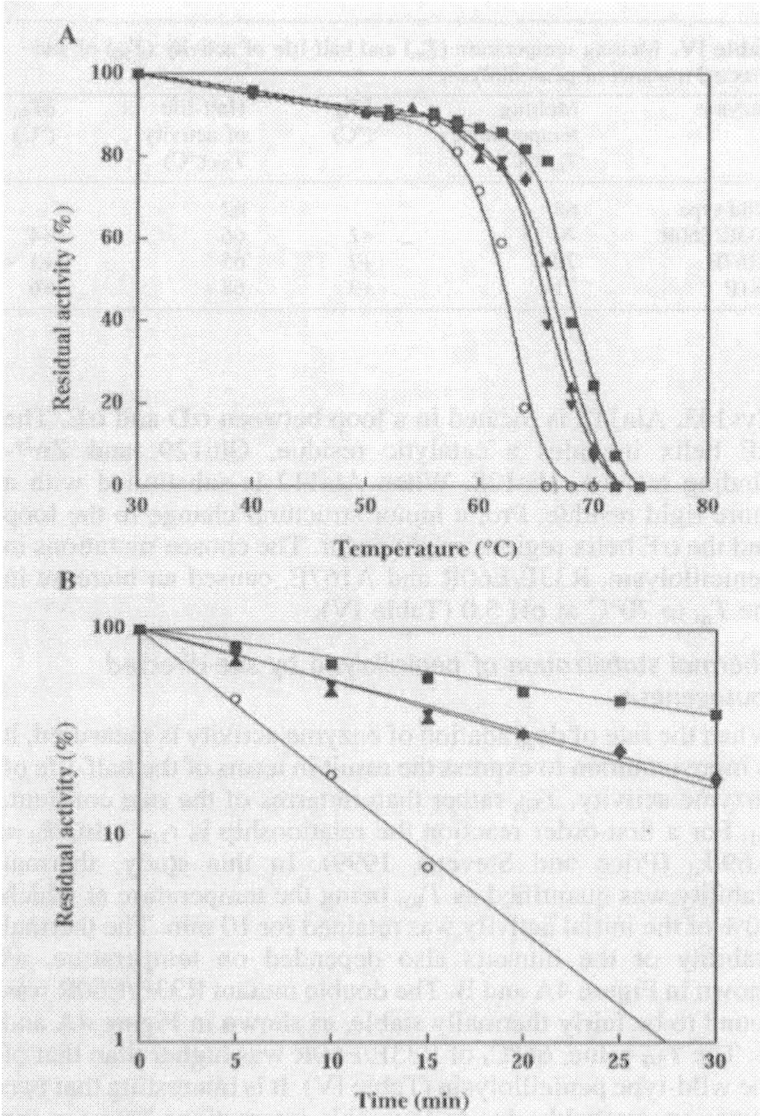

Figure 15. (A) Determination of the half-life of activity, T50, for wild-type penicillolysin (○) and mutants R33E/E60R (▲), A167E (▼) and T81P (■). (B) Stability at 65°C of wild-type of penicillolysin (○), R33E/E60R (▲), A167E (▼) and T81P (■).

(Doi, Y. *et al.*, *Protein Eng. Des. Selection, 17*, 261-266 (2004))

Chapter 6

ACID CARBOXYPEPTIDASE FROM *ASPERGILLUS SAITOI*

6.1. ACID CARBOXYPEPTIDASE

Carboxypeptidase A (EC 3.4.17.1) that cleaves carboxy-terminal aromatic or neutral amino acid from peptides and proteins has been extensively characterized [77], and carboxypeptidase B (EC 3.4.17.2) that cleaves carboxy-terminal arginine or lysine from the peptides and proteins has been studied for many years [78]. Both pancreatic carboxy-peptidases A and B are Zn^{2+}-metalloenzymes with an alkaline pH optimum.

While establishing purification for aspartic proteinase (aspergillopepsin I, EC 3.4.23.18) (7, 8) of *Aspergillus saitoi*, which is a food microorganism strain, it was discovered to be a rich source of acid carboxypeptidase (EC 3.4.16.5), which removes acidic, neutral, and basic amino acids, as well as proline, from the carboxyterminal position at pH around 3 [79, 80]. The optimum pH with Z-Tyr-Leu (Z- = bebzyloxycarbonyl-) of the acid carboxypeptidase from *A. saitoi* was 3.5. The optimum with Z-Glu-Tyr was 3.1, and that with Z-Gly-Pro-Leu-Gly, 3.2.

The acid carboxypeptidase is stable at temperatures of about 50°C for 10 min, but rapid inactivation occurs at temperatures of about 60°C for 10 min.

Competitive inhibition of the carboxypeptidase from *A. saitoi* by small substrates was found with hydrocinnamic acid, indole-3-propionic acid, and 4-phenylbutyric acid [80]. The K_i for hydrocinnamic acid inhibition was 4×10^{-4} M. Diisopropylfluorophosphate (DFP) and tosyl-L-phenylalanylchloromethane (TPCK) were also powerful inhibitors of the carboxypeptidase from *A. oryzae*

[80]. *p*-Chloromercuribenzoate (PCMB) and iodoacetic acid were also powerful inhibitors of the carboxypeptidase from *A. saitoi*, while the inhibitors of DFP, TPCK, PCMB, and iodoacetic acid on the carboxypeptidase from *A. saitoi* were less than that of *A. oryzae* [80]. As the carboxypeptidase activity of *A. saitoi* has no effect when used with ethylenediaminetetraacetate (EDTA) and *o*-phenanthroline, the enzyme is a different type of carboxypeptidase from those of the pancreatic sources, carboxypeptidase A and carboxypeptidase B [80].

Cultivation of fungi of the genus *Aspergillus* is carried out both in solid ('Koji') culture and liquid (submerged) culture. The first extracellular acid carboxypeptidase from *Aspergillus saitoi* was discovered by Ichishima [79]. Using the method of Yphantis, the extrapolated molecular mass was determined to be 139,000 as shown in Figure 16 [79].

According to the gel filtration molecular mass of 155,000 for the larger form or polymer and 51,000 for the smaller form or monomer were obtained in the absence of NaCl, and in the presence of 0.2 M NaCl a value of 135,000 was obtained (Figure 17) [79].

The interdependence of protein surface and hydration of Sephadex has implications for confusion of the molecular mass experiments. The acid carboxypeptidase from *A. saitoi* is a glycoprotein, containing approximately 30% carbohydrate by weight (81, 82). Furthermore, a single protein band with a molecular mass of 72,000 was detected in SDS-PAGE [79]. This indicates that the carboxypeptidase is a homologous dimmer [79].

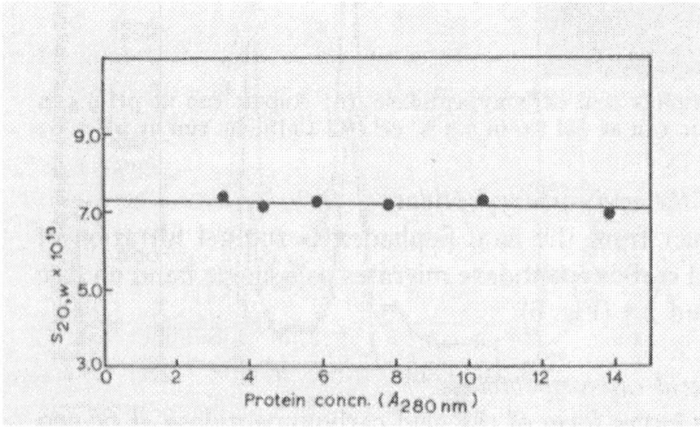

Figure 16. Sedimentation constant, $S^{o}_{20,w}$, of *Aspergillus* acid carboxypeptidase.

(Ichishima, E., *Biochimica. Biophys. Acta, 258*, 274-288 (1972))

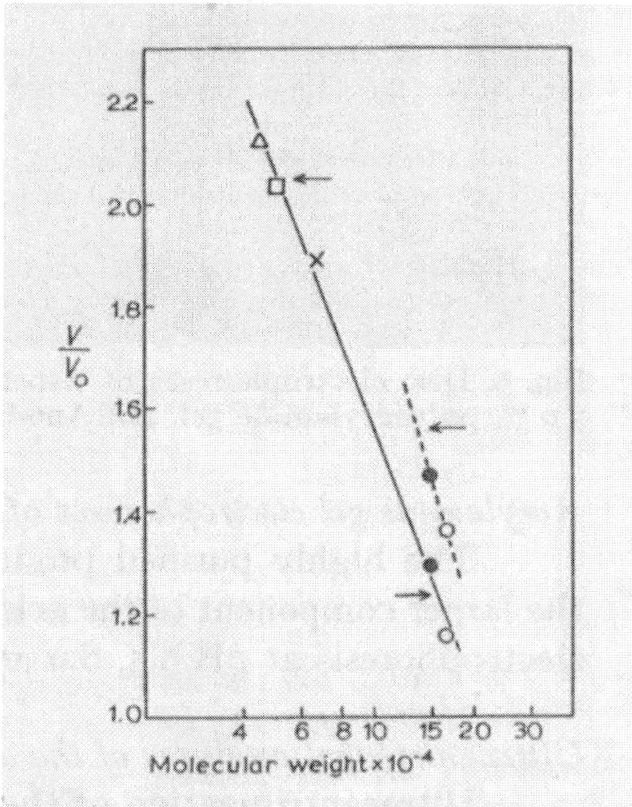

Figure 17. Molecular mass determination of *Aspergillus* acid carboxypeptidase with Sephadex G-200. The column (2 cm x 70 cm) was eluted 0.05 M acetate buffer, pH 3.1, in the absence (—) or presence of 0.2 M NaCl (···). △, egg albumin; □, Taka-amylase A; ×, bovine hemoglobin; ●, bovine γ -globulin; ○, bovine thyroglobulin.

(Ichishima, *Biochim. Biophys. Acta, 258*, 274-288 (1972))

6.2. Specificity of Acid Carboxypeptidase

The relative rates of hydrolysis of a series of peptides by the enzyme are shown in Table 11. The specificity of the acid carboxypeptidase displays the features typical of all pancreatic carboxypeptidases, hydrolysis of the specific substrate R-X-Y between X and Y (R = peptide residue, Z-, Bz-, Ac-). The amino acid in position Y must have a free carboxyl group; dipeptides (having free amino group) are not hydrolyzed. The enzyme hydrolyzes most of the α-amino substituted peptides. The carboxypeptidase was inactive on a number of small amides tried at pH 3.0. A peculiarity of its specificity, however, was its inability to hydrolyze the peptide bond of the tripeptides tried in the Table 11.

Table 11. Comparative rate of hydrolysis for *Aspergillus* acid carboxypeptidase on a range of peptides and amides

The rate of hydrolysis of benzyoxycarbonyl-L-Glu-L-Tyr (Z-Glu-Tyr) is arbitrarily taken to be 100. C-terminal amino acid residues of the substrates are expressed as R-X-Y

Peptides	Relative activity	Peptides	Relative activity
α-Amino-substituted		Tripeptides	
peptides		Gly-Gly-Gly	0
X =aromatic amino		Ala-Gly-Gly	0
acid		Leu-Gly-Gly	0
Z-Tyr-Leu	2502	Gly-Gly-Leu	0
Z-Tyr-D-Leu	0	Dipeptides	
Z-Phe-Leu*	1968	Tyr-Leu	0
Z-Phe-Tyr*	916	Gly-Gly	0
Z-Phe-Tyr-Leu*	1025	Gly-Asp	0
Ac-Phe-TyrI_2*	1	Leu-Gly	0
(2) X = acidic amino		Gly-D-Asp	0
acid		Amides	
Z-Glu-Phe*	184	Z-Tyr-Leu-NH_2*	0
Z-Glu-Tyr	100	Z-Trp-Phe-NH_2*	0
(3) X = leucine		Z-Gly-Phe-NH_2*	0
Z-Gly-Pro-Leu-Gly	30	Z-Gly-Leu-NH_2*	0
(4) X = glycine		Z-Ala-Phe-NH_2*	0
Z-Gly-Leu	4		
Z-Gly-Phe	2		
Z-Gly-Trp	0.2		
Z-Gly-Pro	0		
Z-Gly-Pro-Leu-Gly-Pro	0.2		
Bz-Gly-Lys			
(5) X = valine	1		
Z-Val-Glu			
(6) X = proline	0.4		
Z-Gly-Pro-Leu			
	trace		

*Partially insoluble at pH 3.0.
(Ichishima, E., *Biochim. Biophys. Acta, 56*, 1371-1372 (1992)).

Several qualitative conclusions can be drawn from the results. The acid carboxypeptidase splits off neutral and basic amino acids from the substrates and has the ability to hydrolyze the peptide bond, ---X-Pro in oligopeptides.

Significant enzymatic hydrolysis has been found for substrates in which the penultimate position (X=P_1 site [12]) of the carboxyterminal is occupied by an aromatic or acidic amino acid residue. It will be noted that, for substrates of the type Z-X-Y, a change in the X position from tyrosine to glycine, from phenylalanine to glutamic acid, or from glutamic acid to glycine leads to a marked decrease in the velocity rate. The favorable effect of the aromatic and carboxyl substituents at the side chain of the X residue is emphasized by the finding that where X is an aromatic amino acid residue and glutamic acid, the values of the velocity rate are approximately one-six hundredth and one-tenth that for X = glycine, respectively. It may be concluded, therefore, that with small synthetic substrates of the type Z-X-Y, where the X-Y bond is broken, the acid carboxypeptidase exhibits a preference for aromatic or carboxyl in the X position. Only limited data are available on the effect of changing the X and Y groups on the side chain specificity of the enzyme, but it would appear that this conclusion holds for substrates in which the X group is changed from tyrosine or phenylalanine to glycine.

Thus Z-Tyr-Leu, Z-Glu-Tyr and Z-Gly-Pro-Leu-Gly are good substrates. Z-Phe-Tyr-Leu, Z-Phe-Leu, Z-Phe-Tyr and Z-Glu-Phe were cleaved very easily by the enzyme, however, they are extremely insoluble at the acidic end of the pH range of carboxypeptidase activity (pH 2-4).

The pH dependence of the acid carboxypeptidase from *A. saitoi* catalyzed hydrolysis of Z-Glu-Tyr was determined, and the main conclusions derived from analysis of the data were that two catalytically active groups on the enzyme with $pK_{e1} = 2.3$ and $pK_{e2} = 4.9$ were important in the enzymatic action at pH 3.1 [80]. The pK_{e1} of 2.3 could represent participitation of anionozed carboxyl group, whereas the $pK_{e2} = 4.9$ could be in accordance with the ionization of an imidazole group or a carboxyl group [80].

A. saitoi carboxypeptidase has been used for automatic C-terminal amino acid sequence analyses of anticoagulant decapeptide (SQLQEAPLEK) [83], α-amylase (-LR) from cultured rice (*Oryza sativa*) cells [84], and a recombinant *Serratia marcesence* serine proteinase [85].

The acid carboxypeptidase from *A. saitoi* releases the carboxyterminal phenenylalanine-amide (-Phe-NH_2) from the carboxy-terminal amidated peptides, such as gastrin-related peptide (*t*-amyloxycarbonyl (Aoc)-Trp-Met-Asp-Phe-NH_2, Aoc-WMDF-NH_2) and molluscan cardioexcitatory neuropeptide (Phe-Met-Arg-Phe-NH_2, FMRF-NH_2) [86]. The summarized data are shown in Table 12. When gastrin-related peptide was used as a substrate, the enzyme acted only as a carboxyamidase because of the presence of the hydrophobic amino acid residue, tryptophan, in the P_3 [12] position.

When eledoisin-related peptide was used as a substrate, the enzyme quickly released methionine-amide rather than methionine from the carboxy-terminus of –FIGLM-NH_2. In this case, the P_3 position was occupied by the hydrophobic amino acid isoleucine. Furthermore, when D-amino acid occupied the P_3 position of the substrate, as in enkepharin-amide, the enzyme did not liberate a carboxy-terminal methionine-amide but acted only as amidase for this substrate. It can be concluded that the S_3-P_3 interaction is important in increasing enzyme-substrate affinity and turnover rate for the release of carboxy-terminal amino acid amide from the carboxyterminal amidated peptides [86]. When FMRF-NH_2 was used as the substrate, approximately 16% of the carboxyterminal peptide bond in terminal arginyl-phenylalanine was cleaved. Therefore, the enzyme acted not only as a carboxyamidase but also as an amidase for the carboxy-terminal peptide bond (FMRF-NH_2). The carboxypeptidase from *A. saitoi* did not act as amidase for a P_4-lacking substrate such as Z-X-Y-NH_2. Thus, the P_4 position is important for the enzyme to bind and act as an amidase [86].

Table 12. Hydrophobicity of side chains of amino acid residue adjacent to carboxy-terminal bond to be split by the carboxypeptidase from *A. saitoi*

Hydrophobicity of Side Chains of Amino Acid Residue Adjacent to Carboxy-Terminal Bond to be Split by the Carboxypeptidase from *Aspergillus saitoi*

Substrate	Amino acid residues adjacent to carboxy terminal bond to be split by the enzyme[a] $\frac{S_4\ S_3\ S_2\ S_1\ S'_1}{P_4\ P_3\ P_2\ P_1\ P'_1}$ ↑↓	Hydrophobicities of side chain of amino acid residues[b] Δgt (kcal/mol)	Relative rate of hydrolysis	Recovery of carboxy terminal amino acid residue (%)
Z-Ala–Phe–NH_2	Z–A–F–NH_2	?–0.5–2.5–?	0	0
	Z–A–FNH_2	?–0.5–2.5	0	0
Gastrin-related peptide	–W–M–D–F–NH_2	3.4–1.3–0.5–2.5–?	0	0
	Aoc–W–M–D–FNH_2	?–3.4–1.3–0.5–2.5	1.0	100
FMRF-amide	F–M–R–F–NH_2	2.5–1.3–?–2.5–?	0.84	76
	F–M–R–FNH_2	2.5–1.3–?–2.5	0.16	14
[D-Ala[2], Met[5]]-enkephalinamide	–DA–G–F–M–NH_2	–0.5–0–2.5–1.3–?	1.0	71
	Y–DA–G–F–MNH_2	2.3–0.5–0–2.5–1.3	0	0
Eledoisin-related peptide	–I–G–L–M–NH_2	–2.9–0–1.8–1.3–?	0.35	31
	–F–I–G–L–MNH_2	–2.5–2.9–0–1.8–1.3	0.65	57

(Takeuchi, M. and Ichishima, E., *Agric. Biol. Chem., 53*, 2301-2306 (1989))

Aspergillus acid carboxypeptidase can hydrolyze the amide bond between the amino acid and 7-amino-4-methylcoumarin (AMC) in the peptide-4-methylcoumaryl-7amide (peptide-MCA) such as Pro-Phe-Arg-MCA (PFR-MCA) at pH 4.9 [87]. In the MCA-substrates, the P_3 position was occupied by a hydrophobic amino acid such as proline or valine.

6.3. Cloning and Expression of Acid Carboxypeptidase Gene (*CpdS*)

Screening of *A. saitoi* cDNA library with synthetic oligonucleotide identified five clones by colony hybridization. One of these colonies was sequenced completely and was found to contain a 1812 bp cDNA that encoded a complete nucleotide sequence of the caboxypeptidase [88]. The cDNA sequence contains a single open reading frame of 532 amino acids starting at position 46 and ending at 1617. The sequence around the proposed methionine start codon closely matches the consensus sequence for eukaryotic translational initiation sites and the stop codon is found in the frame at position 22. Codon usage of the coding region leads the choice of C in the third codon position (C : 53%; G : 20%; T : 18%; A : 9%).

The putative mature *A. saitoi* arboxypeptidase consists of 471 amino acids (molecular mass of 52453 Da). The underlined sequences of amino acid residues in Figure 18 correspond to the sequences found by amino acid sequencing analyses of native caboxypeptidase from *A. saitoi*. Hydropathy analysis by the algorithm of Kyte and Doolittle indicates that the NH_2-terminal portion of the polypeptide is hydrophobic, and analysis based on the signal sequence cleavage prediction method of von Heijne indicates possible cleavage sites after the 18th amino acid. Since *A. saitoi* carboxypeptidase is secreted into the medium, the NH_2-terminal sequence is thought to function as a signal sequence.

Comparison of the deduced sequence of *A. saitoi* carboxypeptidase with other known serine carboxypeptidase sequences shows that they share a low degree of similarity: 32% with wheat carboxypeptidase II, 32.3% with malt carboxypeptidase II and 26.2% with yeast carboxypeptidase Y (Figure 19) [88]. However, all of the sequences conserve the catalytic domains (indicated by boxes II to IV in Figure 19) and the domain (box I in the Figure 19) which contains the amino acid residues recognizing the C-terminal carboxylate group of peptide substrates.

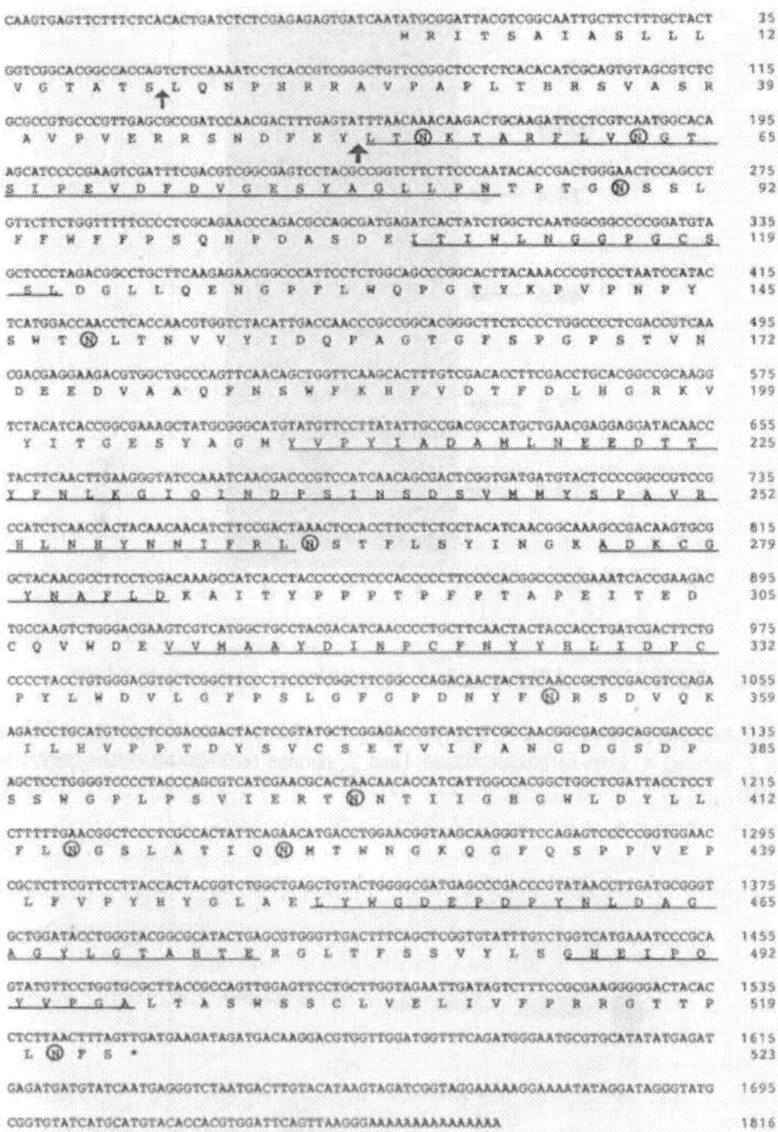

Figure. 18. cDNA sequence of *A. saitoi* carboxypeptidase and deduced amino acid sequence. The predicted amino acid sequence of *A. saitoi* carboypeptidase is shown below the DNA sequence. The standard single-letter symbols are used. Amino acid sequences found by direct sequencing are underlined. A light arrow marks the putative cleavage site of signal peptide according to the weight-matrix approach of von Heijne.

A bold arrow indicates the start of the mature enzyme. The amino acid sequences of the potential N-glycosylation site are circled.

(Chiba, Y. *et al., Biochem. J., 308*, 405-409 (1995))

There are also ten potential sites for N-linked glycosylation present in the sequence.

The *A. saitoi* carboxypeptidase cDNA was cloned downstream of a GDP promoter, and the resulting plasmid, pGCP13, was used to generate recombinant *A. saitoi* carboxypeptidase protein. No enzymatic activity was detected in the culture supernatant. We detected the *A. saitoi* carboxypeptidase activity of the extract obtained from yeast cells transfected with *A. saitoi* carboxypeptidase cDNA in forward orientation (pGCP13), although no activity was observed with the vector alone at pH 3.1.

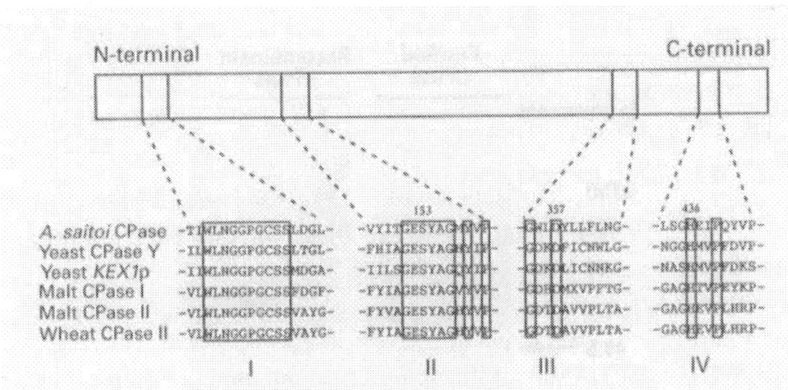

Figure 19. Comparison of the peptide domains reportedly required for enzyme catalysis in eukaryotic serine carboxypeptidases. *A. saitoi* carboxypeptidase (CPase) is aligned with carboxypeptidase Y, *KEX1* protein, malt carboxypeptidase I, malt carboxypeptidase II and wheat carboxypeptidase II sequences with the differences identified. Amino acids conserved with the serine carboxypeptidases sequence are boxed.

(Chiba, Y. *et al., Biochem. J., 308*, 405-409 (1995))

The recombinant *A. saitoi* carboxypeptidase activity is not affected by whether the ZPCK, which is an inhibitor of yeast carboxypeptidase Y, is present or not.

Western-blot analysis of yeast cell extracts shows a 72 kDa protein with rabbit anti-(*A. saitoi* carboxypeptidase) serum, which is consistent with the apparent molecular mass of the native *A. saitoi* carboxypeptidase. Conversely, the extracts obtained from yeast cells transfected with the vector (pG-3) alone or with cDNA in reverse orientation (pGP31), as negative controls, yielded no stainable protein.

The recombinant carboxypeptidase was treated with glycopeptidase F and was subjected to SDS-PAGE. De-N-glycosylated recombinant enzymes migrated with apparent molecular masses of 62 kDa, a little larger than or equivalent to that of de-N-glycosylated carboxypeptidase from *A. saitoi*.

The recombinant carboxypeptidase was purified using gel-filtration and cation-exchange chromatography. The purified recombinant enzyme was homogeneous after SDS-PAGE and had a subunit electrophoretic mobility slightly greater than that of the carboxypeptidase from *A. saitoi*. The recombinant carboxypeptidase was subjected to gel filtration on a TSKgel G3000W XL column in comparison with standard proteins of known size. The estimated molecular mass of the recombinant carboxypeptidase was 135 kDa, which is expected for a dimmer.

The specific activity of the recombinant carboxypeptidase towards Z-Tyr-Leu was measured at pH 3.1. The value for recombinant carboxypeptidase was 4.1 katal/kg, which was almost the same as that of *A. saitoi* carboxypeptidase, 4.5 katal/kg.

As shown in Figure 19, four conserved regions were observed in the *A. saitoi* carboxypeptidase sequence. To establish whether the Ser153, Asp357, and His436 residues in the conserved regions were essential for activity of the *A. saitoi* carboxypeptidase, we generated point mutations that changed each residue to alanine, by oligonucleotide site-directed mutagenesis. The mutated *cpdS* genes were cloned into the pG-3 plasmid and introduced into yeast YPH250 cells. Mutant catboxypeptidases (S153A, D357A, and H436A) were then produced in 5-day cultures. Attempts to detect an inactive protein were made by Western blotting, with the wild-type enzyme as molecular-mass standard. Responses were obtained in all mutants. Next, each extract was assayed with Z-Tyr-Leu at pH 3.1; however, no carboxypeptidase activity was detected from any mutant protein. The Asp138 mutant protein (D138A), as a positive control, completely retained the carboxypeptidase activity. These results suggest that the *A. saitoi* carboxypeptidase functionally conserves the catalytic residues (Ser153, Asp357, and His436) of any other carboxypeptidase and strongly imply that the *A. saitoi* caeboxypeptidase is a serine proteinase with a catalytic triad comprising Asp-His-Ser. It can be concluded that *A. saioti* carboxypeptidase is a member of the serine carboxypeptidase C family (EC 3.4.16.5) [89].

6.4. Debittering of the Bitter Peptides

Enzymatic hydrolysates of various proteins have a bitter taste, which may be one of the main drawbacks to their use in food. Arai *et al.* [90] showed that the bitterness of peptides from soybean protein hydrolysates was reduced by treatment of *Aspergillus* acid carboxypeptidase from *A. saitoi.* Significant amounts of free leucine and phenylalanine were liberated by *Aspergillus* carboxypeptidase from the tetracosapeptide of the peptic hydrolysate of soybean having a bitter taste. Furthermore, the bitter peptide fractions obtained from peptic hydrolysates of casein, fish protein, and soybean protein were treated with wheat carboxypeptidase W [91]. The bitterness of the peptides lessened with an increase in free amino acids. Carboxypeptidase W can eliminate bitter tastes in enzymatic proteins and is commercially available for food processing.

6.5. A New High-Mannose Type N-Linked Oligosaccharide

A saitoi acid carboxypeptidase is a glycoprotein that contains both *N*- and *O*-linked sugar chains [81, 82]. Glycosylation of the *A. saitoi* carboxypeptidase is stabilc against proteinase and proteolysis. In this chapter, the characterization of a new N-linked high-mannose type oligosaccharide from *A. saitoi* acid carboxypeptidase is described.

The sugar chain was cleaved by N-glycanase (Seikagaku Kogyo, Tokyo) from heat-denatured carboxypeptidase with SDS that was purified to homogeneity from the commercial enzyme preparation of *Aspergillus saitoi* ('Molsin', Seisin Pharmaceutical Co., Ltd., Chiba). The cleaved oligosaccharides were purified by gel filtration on a Bio-Gel P 4 column. One major oligosaccharide fraction eluted from the column was used for further structural analyses. The ^{1}H-NMR spectrum of the anomeric proton region of the intact oligosaccharide (Figure 20) was compared with that of the reference oligosaccharide of $Man_9GlcNAc$ [92, 93] from yeast mannoprotein (Table 13). The set of chemical shift values of the intact oligosaccharide was almost identical to that of the yeast $Man_9GlcNAc$ [92, 93], except for the addition of the signal at 5.386 ppm which was assigned to the Man-7 of $Man_{10}GlcNAc_2$ (Figure 20B). Owing to the O-2 substitution of Man-7 in $Man_9GlcNAc_2$ by Man-10, the signals at 5.077 ppm and 5.096 ppm of Man-7 showed a downfield shift (Table 13). Chemical shifts appearing around 4.8 ppm and 5.1 ppm should be assigned to the unsubstituted Man-6 and Ma-7, respectively. There must be some contaminanting sugar chains in $Man_{10}GlcNAc_2$. To confirm the mode of the linkage of terminal mannose, the oligosaccharide was digested with the exo-α1→2-mannosidase from *Aspergillus saitoi* [94, 95] at pH 5.0. The ^{1}H-NMR spectrum of the exo-α1→2-mannosidase resistant fragment indicates the structure identification is that of the standard $Man_5GlcNAc_2$. On the other hand, the ^{1}H-NMR spectrum of core oligosaccharide from yeast mannoprotein earlier digested with exo- 1,2-α-mannosidase gave one additional signal at 4,917 ppm, indicating the structure of $Man_6GlcNAc$ (Table 13).

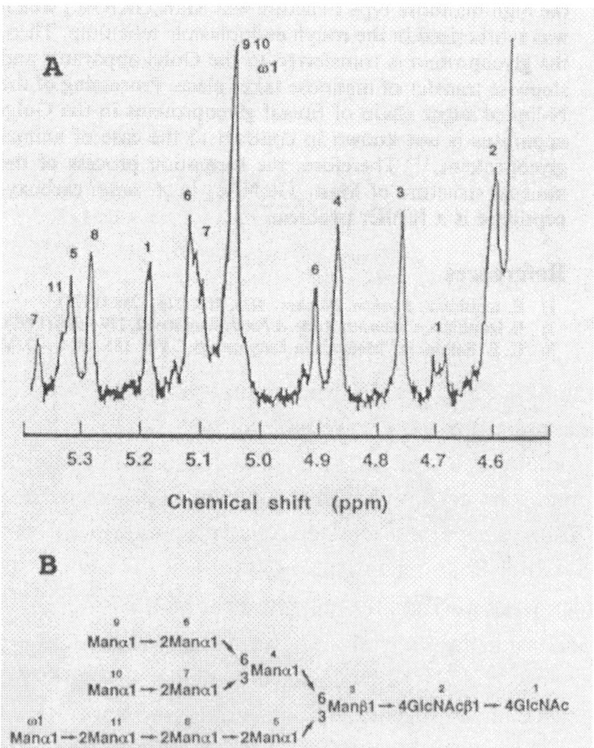

Figure 20. 400 MHz ^{1}H-NMR spectrum of anomeric region at 50°C (A) and proposed structure (B) of the oligosaccharaide from *Aspergillus saitoi* carboxypeptidase. Chemical shift (δ) are expressed in parts per million downfield from internal sodium 2,2-dimethyl-2-silapentane-5-sulfonate, measured by reference to D_2O (δ 4.525 at 50°C).

(Chiba, Y. *et al., Biosci. Biotechnol. Biochem., 56*, 1371-1372 (1992))

The FAB-MS spectrum of the N-linked oligosaccharide from *A. saitoi* carboxypeptidase gave the *m/z* ion as $(M + Na)^+ = 2067$, which is consistent with the molecular mass of $Man_{10}GlcNAc_2$ + 23. The secondary fragments are formed by elimination of Man and Man_2, to give the ions with *m/z* 1905 and 1743, respectively. To analyze the detailed structure of the oligosaccharide, we used partial acetolysis using the pyridylaminated (PA) derivative. Man_5-$GlcNAc_2$-PA, which was derived by partial acetolysis, indicating the oligosaccharide had an extra mannose residue (ω1) substituting for Man-11 in the lower arm of $Man_9GlcNAc_2$. Based upon the above results, the sugar chain structure of *A. saitoi* carboxypeptidase is proposed as Manα1→2Manα→6(Manα1→2Manα1→3)Manα1→6(Manα1→2Manα1→2Manα1→2Manα1→3)Manβ1→4GlcNAcβ1→4GlcNAc (Figure 20 B), which has not been found in yeast glycoproteins [96] or in animal glycoproteins [97].

A number of experiments on the biosynthesis of N-linked oligosaccharides have shown that the maximum length of the high-mannose type structure was $Man_9GlcNAc_2$, which was synthesized in the rough endoplasmic reticulum.

Table 13. ¹H-NMR chemical shifts of oligosacharides from *Aspergillus saitoi* carboxypeptidase and its fragment

Compound	Sugar residue and structure					δ				
	D	C	B	A	A'	D	C	B	A	A
								ppm		
Sugar chain of *A. saitoi* CPase (I)	αM →⁶	αM →⁶	βM →⁴	βGlcNAc →⁴	αβGlcNAc	5.128	4.877	4.767	4.619	5.192 4.693
	↑²	↑³	↑³							
	αM	αM	αM			5.051	5.386	5.328		
		↑²	↑²							
		αM	αM				5.051	5.294		
			↑²							
			αM					5.335		
			↑²							
			αM					5.051		
Exo-α1→2-mannosidase product of I	αM →⁶	αM →⁶	βM →⁴	βGlcNAc →⁴	αβGlcNAc	4.900	4.876	4.762	4.650	5.178 4.699
		↑³	↑³							
		αM	αM				5.098	5.101		
Yeast mannoprotein inner core fragment	αM →⁶	αM →⁶	βM →⁴	αβGlcNAc		5.146	4.876	4.774	5.197 4.675	
	↑²	↑³	↑³							
	αM	αM	αM ←⁶ αM			5.043	5.077 5.096	5.340	4.913	
			↑³							
			αM					5.286		
			↑²							
			αM					5.043		

(Chiba, Y. *et al., Biosci. Biotechnol. Biochem., 56*, 1371-1372 (1992))

Then, the glycoprotein is transferred to the Golgi apparatus and stepwise transfer of mannose takes place. Processing of the N-linked sugar chain of fungal glycoproteins in the Golgi apparatus is not known in contrast to the case of animal glycoproteins [97]. Therefore, the formation process of the unusual structure of $Man_{10}GlcNAc_2$ in *A. saitoi* carboxypeptidase is a further problem.

6.6. Carbohydrate Moiety of Acid Carboxypeptidase

The carbohydrate was determined to be 29.1% of the molecular mass of the carboxypeptidase [82]. Mild alkali treatment of the enzyme, under conditions that effect β-elimination of the sugar-substituted serine or threonine, caused a reduction of the carbohydrate content from 29.1% to 19.3%. This suggests that O-linked sugar chains account for about 10% of the total mass, while another 19% must be N-linked sugar chains. The neutral sugar composition analysis of the hydrolysates of the carboxypeptidase showed only D-mannose, including the presence of high-mannose type sugar chains.

The partially purified N-linked oligosaccharides were pyridylaminated (PA), and the resulting PA-derivatives were analyzed by size-fractionation HPLC (Figure 21). At least eight peaks were detected. The main peaks were eluted at the eluted position of M8.1[3] and M9.1; however, three peaks (terminated fractions M10, M11, and M12 respectively) were eluted behind M9.1. All eight peaks were collected together, and the mixture of the peaks was digested with 1,2-α-D-mannosidase [94, 95] from *A. saitoi*. Digestion of the PA-derivatives with1,2-α-D-mannosidase gave one peak corresponding to the elution position of M5.1, indicating that all of the N-linked oligosaccharides from *A. saitoi* acid carboxypeptidase had M5.1 as a "core" fragment, and several extra mannoses were attached to the M5.1 core through α-1,2 linkages.

The structure of the PA-oligosaccharide derived from the fraction M10 was analyzed previously [81]. In this study, we purified the PA-oligosaccharide derived from the fraction M11 by HPLC, using reverse phase chromatography column (Cosmosil 5C18-P). One major peak (termed fraction 11-C) and two minor peaks (fraction 11-A and B) of PA-oligosaccharides were obtained from fraction M11. The small yield of fractions 11-A and B were not analyzed further. Fraction 11-C was further purified by the size-fractionation HPLC. To analyze the mannosyl linkage of fraction M11, the sequential exoglycosidases digestions were carried out. Digestion of fraction 11-C by 1,2-α-D-mannosidase (*A. saitoi*) [94, 95] at pH 5.0 yielded a new peak corresponding to the elution position of M5.1. This product was successively digested with α-mannosidase (Jack Bean), then four mannose residues were released. These results indicate that fraction 11-C has four mannose residues with α-1,2- or α-1,6-linkages and another six mannose residues with α-1,2-linkages located outside of the core.

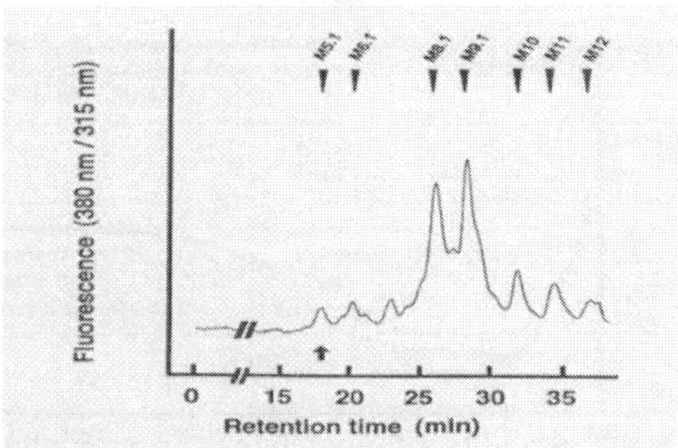

Figure 21. Elution patterns of PA-oligosaccharides derived from *A. saioti* acid carboxypeptidease. Arrow (↑) indicate the elution position of the derivatives after digestion with 1,2- α -D-mannosidase.

(Chiba, Y. *et al., Curr. Microbiol., 27*, 281-288 (1993))

Partial acetolysis of the fraction 11-C under conditions in which α-1,6-mannosidic linkages are cleaved preferentially gave $Man_6GlcNAc_2$-PA. The $Man_6GlcNAc_2$-PA was digested with 1,2-α-D-mannosidase (*A. saitoi*) [94, 95], resulting in a new product which was eluted at the elution position of M2. It suggested that the lower arm of this oligosaccharide had a structure as follows; Manα1-2Manα1-2Manα1-2Manα1-2Manα1-3Manβ1-4GlcNAcβ1-4GlcNAc. From these results, the structure of Fraction 11-C was proposed to be M11 ($Man_{11}GlcNAc_2$) as shown in Figure 22.

The O-linked sugar chains were prepared by alkaline treatment from the acid carboxypeptidase, and β-elimination products were separated from the polypeptide by precipitation with 7% ethanol. The liberated oligosaccharides (70% ethanol-soluble fraction) were analyzed by HPLC. Only one peak was found at a position corresponding to a monosaccharide (Figure 23). On paperchromatographic analysis, only one spot was also obtained at the position of mannitol. The O-linked sugar chains were also analyzed by gas chromatography as alditol trifluoroacetate. Only one peak was detected as a mannitol derivative (Figure 23 A). These results indicate the O-linked sugars of *A. saitoi* acid carboxypeptidase are composed of single mannose residue.

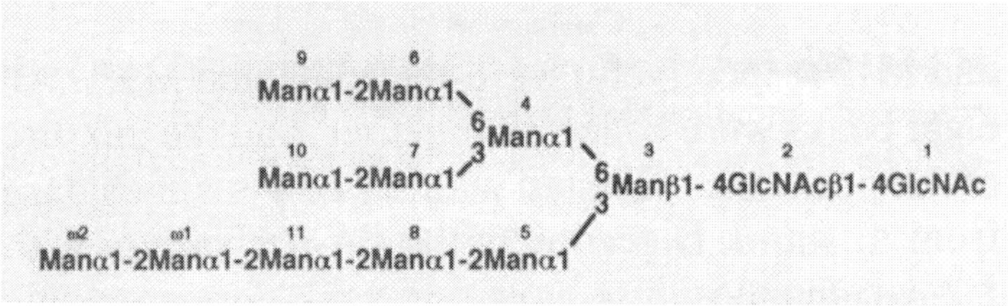

Figure 22. Proposed structure of N-linked oligosaccharide of the fraction 11-C from *A. saitoi* acid carboxypeptidase.

(Chiba, Y. *et al., Curr. Microbil., 27*, 281-288 (1993))

Incubation of the purified acid carboxypeptidase with Endo H and α-mannosidase (Jack Bean) effected a decrease in its molecular mass and in the width of the protein band on SDS-PAGE. Molecular masses of 136 kDa (dimeric form) and 72 kDa (monomeric form) were estimated for the native (glycosylated) acid carboxypeptidase by low-angle laser lightscattering photoenergy and SDS-PAGE, respectively. The deglycosylated enzyme migrated under identical conditions with an apparent molecular mass of 129 kDa (dimmer) and 60 kDa (monomer), respectively.

To investigate whether the oligosaccharide chain influenced protein conformation, the CD and absorption spectra of glycosylated and deglycosylated forms were measured. Slight differences were found between them (Figure 24). From the CD spectra, the contents of α-helix, β-structure, and random coil were calculated (Table 14). The contents of α-helix in the enzymes were almost the same, whereas the contents of β-structure increased in the deglycosylated enzyme. The absorption spectra of the two forms showed no apparent changes. The specific activity of the enzyme decreased slightly from 0.32 katal/kg to 0.31 katal/kg upon deglycosylation. The K_m and k_{cat} values of the native and the deglycosylated enzymes showed no differences

(Table 15). The optimal pH and temperature were unaffected by the deglycosylation process. Both forms of the acid carboxypeptidases were most active at pH 3.0-3.5 towards Z-Glu-Tyr. Native acid carboxypeptidase was fully active below 50°C but gradually became inactivated above 55°C and denatured completely at 70°C.

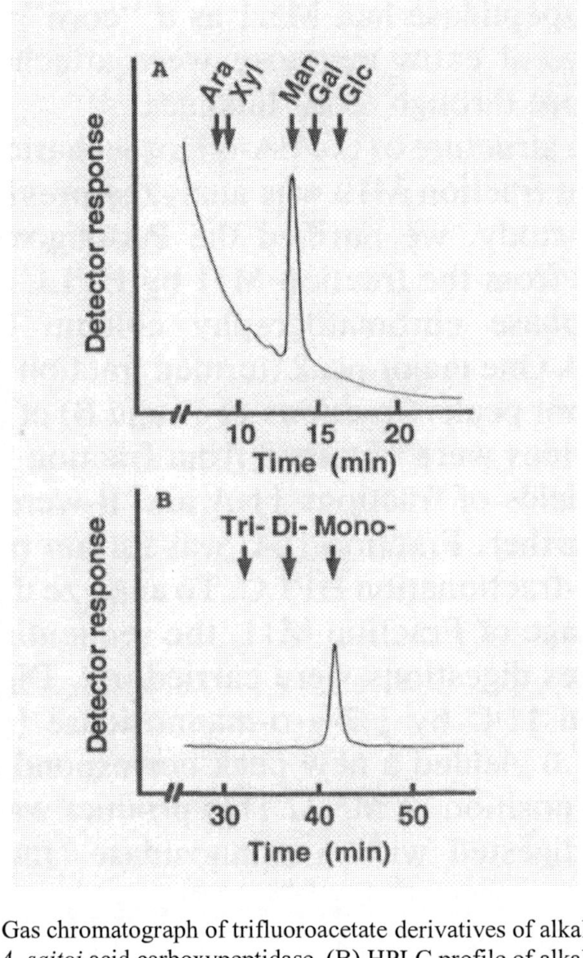

Figrue 23. (A) Gas chromatograph of trifluoroacetate derivatives of alkali-treatment products from *A. saitoi* acid carboxypeptidase. (B) HPLC profile of alkali-treatment products; Mono-, monosaccharide; Di-, disaccharide; Tri-, trisaccharide.

(Chiba, Y. *et al., Curr. Microbiol., 27*, 281-288 (1993))

Table 14. Secondary structures of native and deglycosylated acid carboxypeptidases from *Aspergillus saitoi*

Enzyme	α-Helix (%)	β-Sheet (%)	β-Turn (%)	Random (%)
Deglycosylated Carboxypeptidase	21.2	30.5	21.8	26.5
Native carboxypeptidase	19.1	23.5	27.1	30.3

(Chiba, Y. *et al., Curr. Microbiol., 27*, 281-288 (1993))

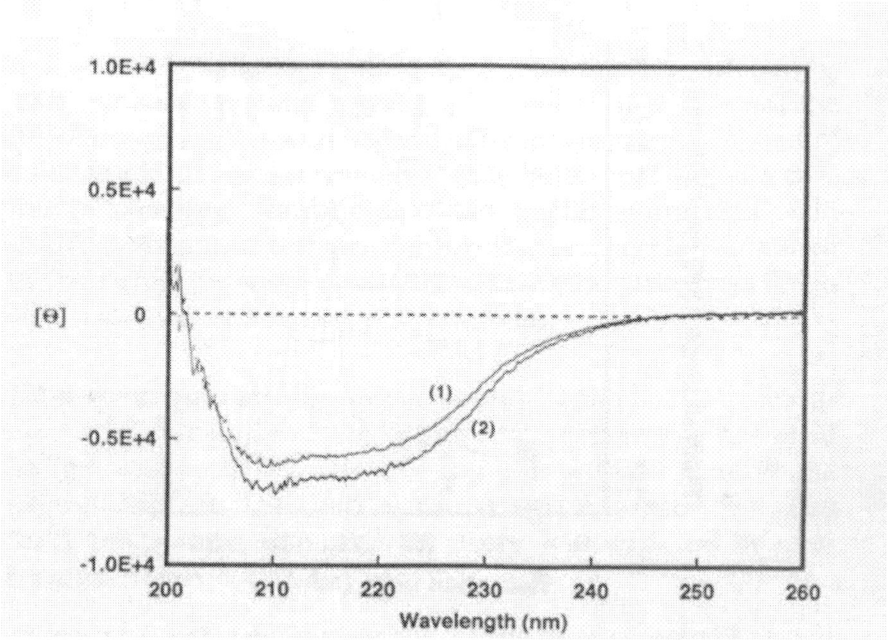

Figure 24. Molecular ellipticity [θ] of native and deglycosylated *A. saitoi* acid carboxypeptidase. (1) Native carboxypeptidase. (2) Deglycosylated carboxypeptidase.

(Chiba, Y. *et al., Curr. Microbiol., 27*, 281-288 (1993))

The temperature activity profile of the deglycosylated enzyme showed a pattern similar to that of the native enzyme.

Deglycosylation had no marked effect on the stability of the acid carboxypeptidase. Both native and deglycosylated enzymes were stable up to 50°C and active in the presence of 4 M urea after 24 h. The molecular structures of both enzymes were not impaired by pepsin at pH 2.5 (Figure 25). These two enzymes had similarly great resistance to peptic digestion after 24h.

Table 15. Kinetic parameters of active and deglycosylated acid carboxypeptidases from *Aspergillus saitoi* toward Z-Glu-Tyr at pH 3.1 and 30°C

Enzyme	K_m (mM)	k_{cat} (s^{-1})	k_{cat}/K_m ($s^{-1} \cdot mM^{-1}$)
Deglycosylated carboxypeptidase	3.8	97.6	25.7
Native carboxypeptidase	3.6	93.6	26.0

(Chiba, Y. *et al., Curr. Microbiol., 27*, 281-288 (1993)).

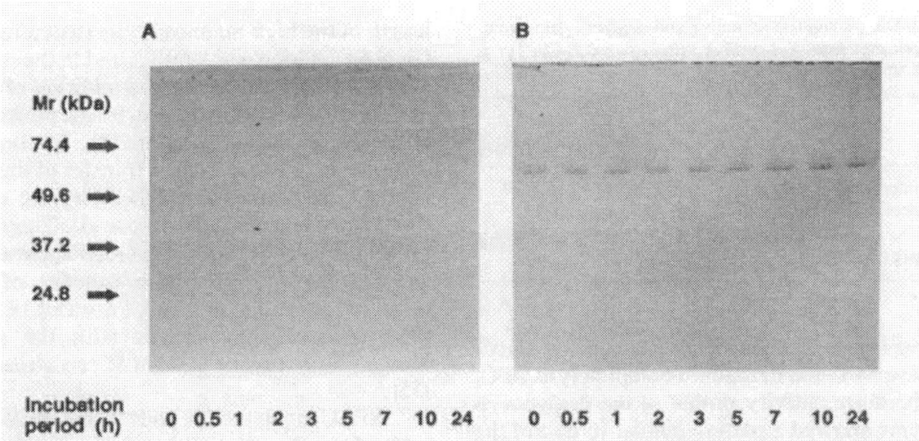

Figure 25. Effect of pepsin treatment on the molecular weight of (A) native and (B) deglycosylated *A. saitoi* acid carboxypeptidase.

(Chiba, Y. *et al., Curr. Microbiol., 27*, 281-288 (1993))

The mycelial wet weight and the protein contents of the mycerial extracts and of the culture filtrates were determined at intervals during growth of mycelia. Addition of tunicamycin (final concentration 5 μg/ml) to the culture medium at the beginning of the culture period had little effect on the growth of the fungus, and the protein concentration of the culture filtrate was slightly decreased. On the other hand, a marked decrease of the acid carboxypeptidase activity in the culture filtrate was found (Figure 26 A).

To examine whether unglycosylated acid carboxypeptidase can be secreted into the medium, we assayed the medium in the presence or absence of tunicamycin by Western blotting analysis. The results indicate that the acid carboxypeptidase was not secreted into the culture medium when tunicamycin was added (Figure 26 B). This suggests that the decrease of enzymatic activity in culture filtrates of mycelia grown in the presence of tunicamycin is largely owing to the decrease of the concentration of the enzyme.

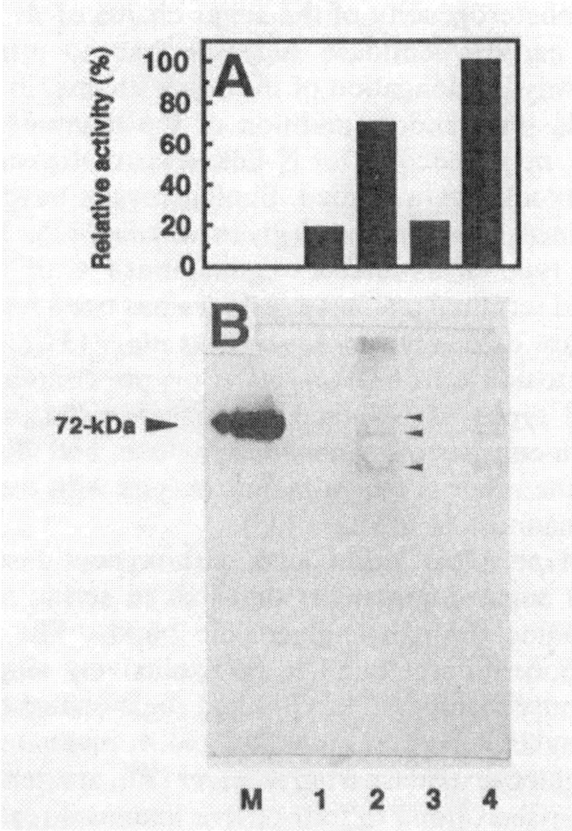

Figure 26. (A) Relative activity of *A. saitoi* acid carboxypeptidase produced in the culture filtrate in the presence and absence of tunicamycin. (B) Western blot analysis of the culture filtrate against the anti-acid carboxypeptidase antibody on SDS-PAGE: (1) with tunicamycin, cultured after 60 h. (2) without tunicamycin, cultured after 60 h. (3) with tunicamycin, cultured after 72 h. (4) without tunicamycin, cultured after 72 h. (M) Purified acid carboxypeptidase.

(Chiba, Y. *et al., Curr. Microbiol., 27*, 281-288 (1993))

Chapter 7

1,2-α-D-MANNOSIDASE FROM *ASPERGILLUS SAITOI*

7.1. 1,2-α-D-MANNOSIDASE

The importance of α-mannosidase (EC 3.2.1.24) in the processing system of glycoproteins in higher eukaryotic cells is well known [98-101]. In mammalian cells, 1,2-α-mannosidases are an essential in the early steps of N-linked oligosaccharaide maturation [102]. In the Enzyme Nomenclature Recommendations [103], three types of α-mannosidases, namely α-D-mannoside mannohydrolase (EC 3.2.1.24), 1,2-α-mannosyl-oligosaccharide-α-D-mannohydrolase (EC 3.2.1.1.113) and 1,3-(1,6)-mannosyl-oligosaccharide α-D-mannohydrolyase (EC 3.2.1.114), are given as systematic names.

We have described the purification to homogeneity of 1,2-α-D-mannosidases from fungi, *Aspergillus saioti* (*A. phoenicis*) [94, 95], *Penicillium citrinum* [104, 105] and *Pycnoporus coccineus* [106], specifically capable of releasing α(1-2)-linked mannose residues from oligosaccharides derived from asparagines-linked glycans of glycoproteins. Ballou [17] isolated and characterized the exo-1,2-α-mannosidase from *A. saitoi* and examined its properties.

The predicted amino acid sequence of the 1,2-α-D-mannosidase from *A. saitoi* [107] is 70, 26, and 35%, identical to those of *P. citrunum* 1,2-α-D-mannosidase [108], yeast (*S. cerevisiae*) Man_9-specific mannosidase [109, 110] and mouse Golgi 1,2-α-D-mannosidase [111], respectively.

7.2. Expression of *A. saitoi* 1,2-α-D-Mannosidase Gene (*msdS*) in *A. oryzae* Cells

For the construction of an overexpression system of the intracellular 1,2-α-mannosidase (EC 3.2.1.113) gene (*msdS*) from *A. saitoi* (*A. phonenicis*), the NH_2-terminal signal sequence of the gene was replaced with that of the aspergillopepsin I (EC 3.4.23.18) gene (*apnS*) signal [112]. Then the fused 1,2-α-mannosidase gene (f-*msdS*) was inserted into the *Not*I site between P-No8142 and T-*agdA* in the plasmid pNAN 8142 (9.5 kbp) and thus the *Aspergillus oryzae* expression plasmid pNAN-AM1 (11.2 kbp) was constructed. The fused f-*msdS* gene has been overexpressed in a transformant *A. oryzae niaD* AM1 cell. The recombinant enzyme expressed in *A. oryzae* cells was purified to homogeneity in two steps. The system is capable of making as much as 320 mg of the enzyme/liter of culture.

The recombinant enzyme has activity with methyl-2-*O*-α-D-mannopyranosyl α-D-mannopyranoside (Man-α1,2-Man-OMe) at pH 5.0, while no activity was determined with methyl-3-*O*-α-D-mannopyranosyl α-D-mannopyranoside (Man-α1,3-Man-OMe) or methyl-6-*O*-α-D-mannopyranosyl α-D-mannopyranoside (Man-α1,6-Man-OMe). The substrate specificity of the enzyme was analyzed by using pyridylaminated (PA)-oligomannose-type sugar chains, $Man_{9\text{-}6}(GlcNAc)_2$-PA (Man is mannose; GlcNAc is N-acetylglucosamine). The enzyme hydrolysed $Man_8GlcNAc_2$-PA (type 'M8A', Manα1-2Manα1-2Manα1-3[Manα1-3(Manα1-2Manα1-6)Manα1-6]Manβ1-4GlcNAcβ1-4GlcNAc-PA) fastest. Meanwhile, the results showed that the poorest substrate among the PA-sugar chains was 'M6C' (Manα1-3[Manα1-2Manα1-3(Manα1-6)Manα1-6]Manβ1-4GlcNAcβ1-4GlcNAc-PA) for the *A. saitoi* 1,2-α-D-mannosidase, and it was assumed that the Manα1-2 residue in M6C interfered with the hydrolysis of other Manα1-2 residues.

Molecular mass values of the enzyme were determined to be 63 kDa by SDS/PAGE and 65 kDa by gel filtration on Superose 12 respectively. The p*I* value of the enzyme was 4.6. The NH_2-terminal amino acid sequence of the enzyme was GSTQSRADAIKAAFSHAWDGYLQY, and sequence analysis indicated that the signal peptide from the aspergillopepsin I gene (*apnS*) was removed. The molar absorption coefficient, ε, at 280 nm was determined as 91539 $M^{-1} \cdot cm^{-1}$. Contents of the secondary structure (α-helix, β-structure and the remainder of the enzyme) by far-UV CD determination were about 55, 38 and 7%, respectively. The melting temperature, T_m, of the enzyme was 71°C by differential scanning calorimetry. The calorimetric enthalpy, ΔH_{cal}, of the enzyme was calculated as 13.3 kJ • kg of protein^{-1}.

7.3. Catalytic Residues of Ca^{2+}-Independent 1,2-α-D-Mannosidas from *A. Saitoi*

Preliminary site-directed mutagenesis experiments of the fused 1,2-α-D-mannosidase gene (f-*msdS*) on the expression vector, pGAM-1, were done previously to identify the functional role of catalytic residues in the 1,2-α-D-mannosidase from *A. saitoi* expressed in yeast cells [113].

Table 16. Specific activities of purified wild type and mutant 1,2-α-D-mannosidases from *A. saitoi*

Mutant 1,2-α-D-mannosidase	Specific activity (mkatal/kg[a])	Relarive activity (%)
Wild type	47.3	100
E124Q	0.009	0.020
E124D	0.096	0.20
D269N	ND[b]	
D269E	0.92	1.9
E273Q	0.007	0.015
E273D	0.30	0.63
E411Q	ND[b]	
E411D	0.35	0.74
E414Q	0.004	0.009
E414D	0.19	0.40
E474Q	0.003	0.006
E474D	0.25	0.53
E504Q	7.0	15

[a]Using Manα1,2Man-OMe as substrate. One katal releases 1 mol of reducing group (calculated as mannose) per second at 30°C, pH 5.0.

[b]ND, not detected.

(Tatara, Y. *et al., J.Bboil.Chem., 278*, 25289-25294 (2003))

Seven carboxylic acid residues to be studied were constructed in the cloned *A. saitoi* 1,2-α-D-mannosidase gene (*msdS*) by site-directed mutagenesis. The roles of six conserved active carboxylic acids in the catalytic mechanism of *A. saitoi* 1,2-α-D-mannosidase were identified by site-directed mutagenesis and kinetic analyses [114]. Recombinant enzymes were purified homogeneously on SDS-PAGE gels (Figure 27A) and identified as *A. saitoi*

1,2-α-D-mannosidase by Western blot analysis (Figure 27B). Purified forms of all recombinant enzymes were assayed by the Somogyi-Nelson method with Man-α1,2-Man-OMe as a substrate. No activity was detected under standard assay conditions for any of the mutant enzymes except E504Q, and the specific activity could only be determined after extended incubation, typically at 60-90 min (Table 16).

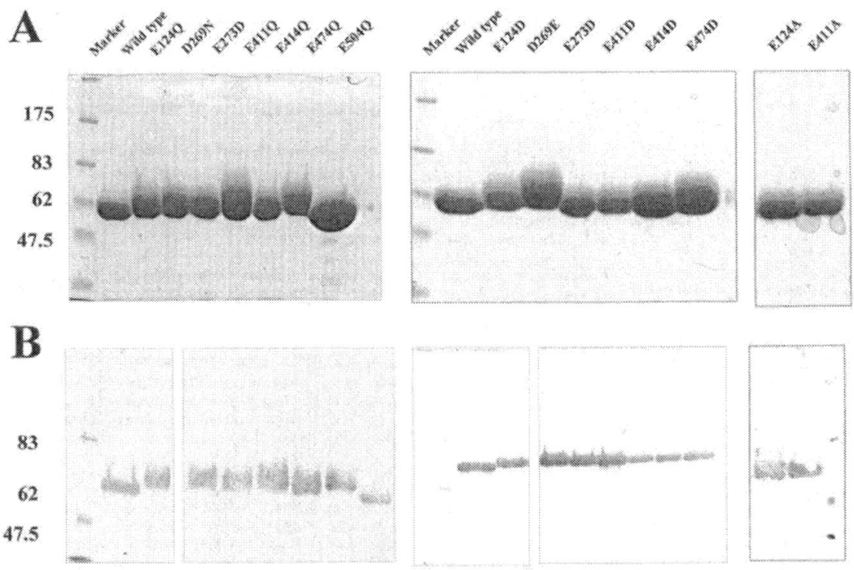

Figure 27. SDS-PAGE (A) and Western blot analysis (B) of the *A. saitoi* 1,2-α-D-mannosidase and mutants.

(Tatara, Y. *et al., J. Biol. Chem., 278*, 25289-25294 (2003))

From Table 16, it appears obvious that the mutations generated resulted in enzymes with significantly reduced ability to hydrolyze Man-α1,2-Man-OMe, as their specific activity is at least 10^3-fold less than that of the wild type enzyme. The specific activities of D269N and E411Q mutants in particular were not detected completely. The results of several experiments suggest that there is no significant contamination by wild type 1,2-α-D-mannosidase in the low activity of mutant enzymes (D269N and E411Q).

pH-activity profiles of E124A and E124D mutant 1,2-α-D-mannosidases for Manα1,2-Man-OMe hydrolysis are shown in Figure 28. The optimum pH of E124D and E124A mutant enzymes shifted from 5.0 to 4.0 and 4.5, respectively. The activity of E124A mutant enzyme was found to be lower than that of wild type enzyme over the whole pH range studied, but the effect was more pronounced for the basic limb.

Table 17. Kinetic parameters of wild type and mutant 1,2-*α*-D-mannosidases from *A. saitoi*

Mutant 1,2-α-D-mannosidase	K_m (mM)	k_{cat} (s^{-1})	k_{cat}/K_m ($s^{-1} \cdot mM^{-1}$)	(%)
Wild type	0.28	2.5	9.04	100
E124Q	0.13	0.0078	0.062	0.7
E124D	0.49	0.022	0.044	0.49
D269N	ND[a]	ND[a]	ND[a]	
D269E	4.8	0.51	0.11	1.2
E273D	7.0	0.37	0.054	0.6
E411Q	ND[a]	ND[a]	ND[a]	
E411D	8.0	1.4	0.18	2.0
E414D	8.6	0.037	0.0043	0.05
E474D	6.4	0.023	0.0036	0.04

[a]ND, not detected.

(Tatara, Y. *et al., J. Biol. Chem., 278*, 25289-25294 (2003))

Kinetic characterization of E124Q, E124D, D269E, E273D, E411D, E414D, and E474D mutants and wild type 1,2-α-D-mannosidase was perfomed at pH 5.0 using Man-α1,2-Man-OMe as a substrate. The results of these kinetic experiments are in Table 17. Kinetic parameters for D269N and E411Q mutants could not be measured since the reaction rates were below the detectable limit. The k_{cat} value for E124Q mutant was 0.0078 s^{-1}, almost 0.3% that of the wild type enzyme. E124D mutant enzyme also decreased the k_{cat} value (0.9% of wild type enzyme). The K_m values of E124Q and E124D mutant enzymes were essentially unchanged. Compared with wild type 1,2-α-D-mannosidase, the D269E and E411D mutant enzymes showed 17- and 28-fold increases in K_m values, respectively, whereas the k_{cat} values were slightly lower than that of wild type enzyme. The K_m values of E273D, E414D, and E474D mutant enzymes were greatly increased to 23-31 -fold that of wild type enzyme, and the k_{cat} values were decreased (1-15% wild type enzyme).

The effects of an external nucleophile on alanine mutant enzymes were studied. Mutation (Glu or Asp to Ala) at the general base catalyst creates a cavity in the active site that can accommodate a small external nucleophile. Sodium salts of azide at 3 M were used as external nucleophiles in the enzyme-catalyzed hydrolysis of $Man_6GlcNAc_2$-PA sugar chain. The reaction mixture was analyzed by HPLC (Figure 29).

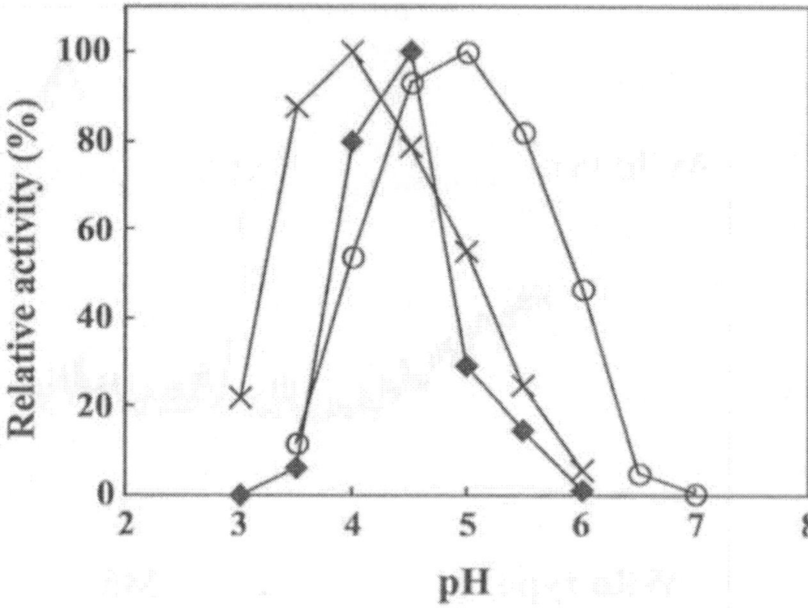

Figure 28. pH-activity profiles of wild type *A. saitoi* 1,2-α-D-mannosidase (○), and mutant E124D (×) and E124A (◆) mutant enzymes represented as relative activities.

(Tatara, Y. *et al., J. Biol. Chem., 278*, 25289-25294 (2003))

The activity of wild type 1,2-α-D-mannosidase was inhibited by a high concentration of azide. The E124A, D269N, and E411A mutant enzymes were inactive in the absence of azide, as a peak corresponding to a $Man_5GlcNAc_2$-PA was not detected. In the presence of azide, these mutant enzymes could not be reactivated.

Atomic absorption spectrophotometric analysis and kinetic analysis of wild-type 1,2-α-D-mannosidase were performed. Concentrations of Ca^{2+} were determined by atomic absorption spectrophotometric analysis. The purified recombinant wild type 1,2-α-D-mannosidase almost completely did not contain Ca^{2+}. Other divalent metal cations including Mg^{2+}, Mn^{2+}, Co^{2+}, Cu^{2+}, and Zn^{2+} were also not detected. After Ca^{2+} treatment (see "Experimental Procedures" of references (114)), Ca^{2+} content was determined as 0.9 mol/mol of wild type 1,2-α-D-mannosidase. Kinetic parameters of Ca^{2+}-free and Ca^{2+} binding 1,2-α-D-mannosidase were determined (Table 18); the k_{cat} and K_m values were little affected by the Ca^{2+} binding.

EDTA inhibited the Ca^{2+}-free wild type 1,2-α-D-mannosidase. The inhibition was investigated by fixed concentrations of EDTA in the assay buffer and by varying the concentrations of substrate. A Lineweaver-Burk plot showed a pattern consistent with competitive inhibition (Figure 30) and gave a K_i of 0.91 mM for EDTA.

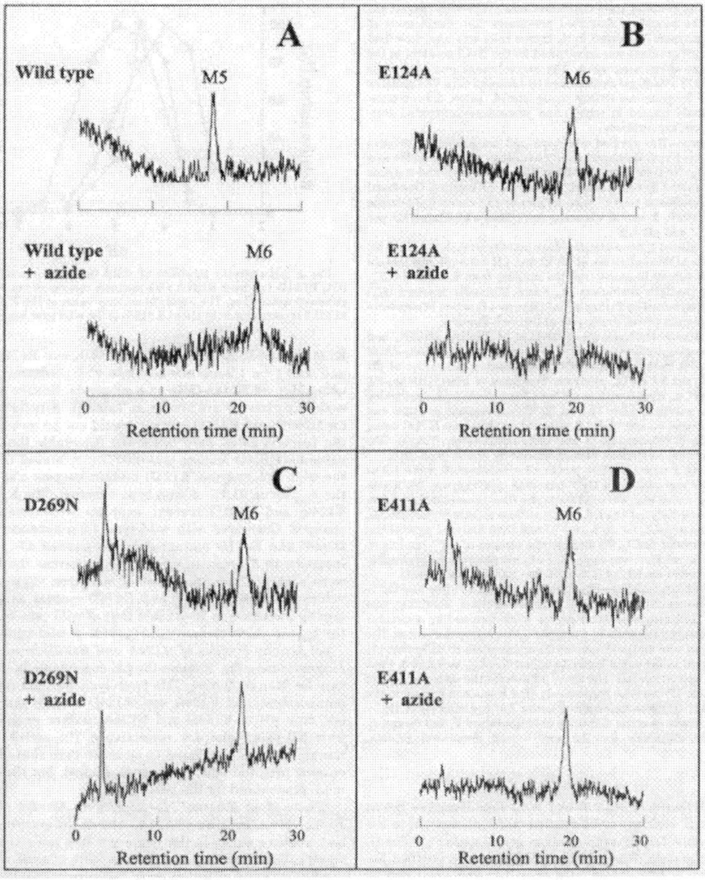

Figure 29. Effect of an external nucleophile on catalytic residue mutants of *A. saitoi* 1,2- α -D-mannosidase using $Man_6GlcNAc_2$-PA as a substrate. A mixture (20 μ l) of 10 pmol Man_6GlucNAc-PA, 3 M sodium azide, and a mutant or the wild type enzyme was incubated at pH 5.0 (50 mM acetate buffer) and 30°C for 72 h. A, 0.25 μ M wild type enzyme. B, 9.7 μ M E124A mutant enzyme. C, 5.9 μ M D269N mutant enzyme. D, 10.5 μ M E411A mutant enzyme.

(Tatara, Y. *et al., J. Biol. Chem., 278*, 25289-25294 (2003))

For Ca^{2+} binding properties of mutant 1,2-α-D-mannosidase, Ca^{2+} concentration of the mutant 1,2-α-D-mannosidase was determined by atomic absorption spectrophotometry after Ca^{2+} binding treatment. As shown in Table 18, the Ca^{2+}-free wild type enzyme bound 0.9 mol of calcium/mol of protein, ~1 mol/mol. E124D, E411D, and E504Q mutants also contained 0.8, 1.1, and 0.9 mol of calcium/mol of protein, respectively. E124Q, D269E, E269N, E474D contained less calcium than 1 mol of calcium/mol of protein. E411Q, E273Q, E273D, E414Q, and E414D mutants demonstrated an almost complete loss of binding of Ca^{2+}.

Table 18. Kinetic parameters of wild type and Ca^{2+} biding 1,2-α-D-mannosidase from *A. saitoi*

1,2-α-D-mannosidase	Ca^{2+} binding	K_m	k_{cat}	k_{cat}/K_m
	(mol/mol)	(mM)	(s^{-1})	($s^{-1} \cdot mM^{-1}$)
Wild type	0	0.28	2.53	9.04
Ca^{2+}-treated[a]	0.9	0.29	1.95	6.84

[a]The Ca^{2+}-free wild type enzyme was treated with $CaCl_2$ as described under "Experimental Procedures" of references [114].

(Tatara, Y. *et al., J. Biol. Chem., 278*, 25289-25294 (2003))

Effects of Ca^{2+} binding on 1,2-α-D-mannosidase thermal stability were determined. The spectra of Ca^{2+}-free and Ca^{2+}-bound 1,2-α-D-mannosidase at far and near UV region were consistent with each other. The Ca^{2+} binding did not undergo a large conformational change. Thermally induced loss of secondary and tertiary structure was monitored at 222 and 291 nm, respectively (Figure 31). T_m, the temperature at which half of the molecules are unfolded, for Ca^{2+}-free 1,2-α-D-mannosidase was 56°C and 58°C at 222 and 291 nm, respectively. The Ca^{2+} binding contributed to the thermal stability increasing T_m to 58 °C and 62°C at 222 and 291 nm, respectively.

For identification of catalytic amino acid residues, drastic reductions in the values of the catalytic coefficient (k_{cat}) were observed in the E124Q and E124D mutant enzymes (Table 17). This may result from significant impairment of the chemical steps in the catalytic reaction. The K_m value was essentially unchanged from wild type enzyme. It is, thus, clear that Glu124 participates in glucosidic bond hydrolysis. E124D shifted the optimum pH to be more acidic (Figure 28). This result is shown the lower pK_a of Asp than that of Glu and supports the assumption that Glu124 is directly involved in the catalytic mechanism. It could not be determined whether the Glu124 residue is a carboxylate group (COO^-) or carboxyl group (COOH). The pH dependence of E124A affected the basic limb, which was generally assumed to be due to the general acid catalyst. An external nucleophile sodium azide did not reactivate the E124A mutant enzyme (Figure 28). Thus the Glu124 may act as an acid catalyst. The mechanism of the all carbohydrate hydrolases usually involves a pair of carboxylic acid residues. In *Penicillium* 1,2-α-D-mannosidase, two of the three residues (Glu124, Asp269, and Glu411 in *A. saitoi* enzyme) are believed to be potentially catalytic residues [115]. Glu124 was suggested to act as an acid catalyst as discussed above. The most likely candidates for other catalytic residues are Asp269 and Glu411. However, D269E and E411D did not show a considerable decrease in the k_{cat} values (Table 17).

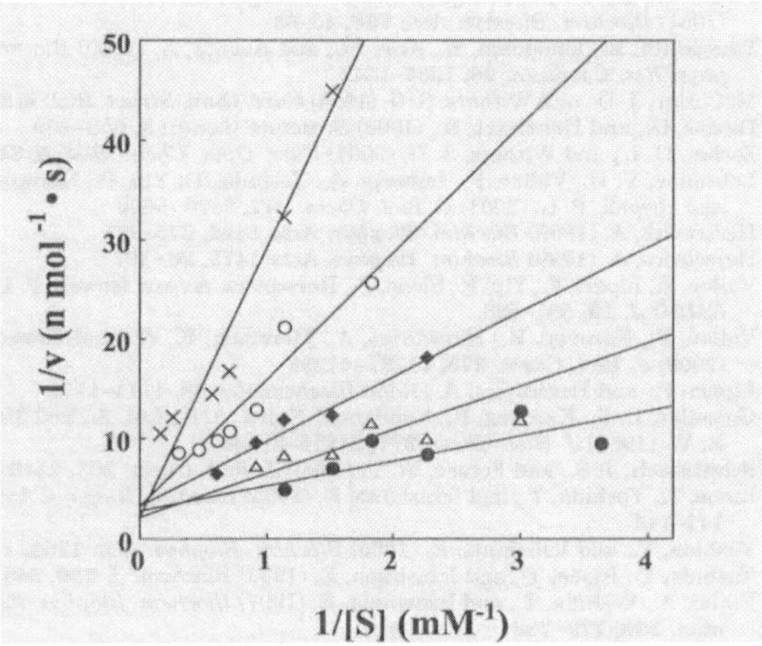

Figure 30. Lineweaver-Burk plot of *A. saitoi* 1,2-α-D-mannosidase in the presence of EDTA. The Ca^{2+}-free wild type 1,2- α -D-mannosidase was assayed at the indicated concentration of EDTA. ●, none; △, 0.5 mM; ◆, 1 mM; ○, 2 mM; ×, 5 mM.

(Tatara, Y. *et al., J. Biol. Chem., 278*, 25289-25294 (2003))

Recently, a base catalyst of an inverting β-amylase (EC 3.2.1.2) from *Bacillus cereus* var. *mycoides* was determined by the chemical rescue methodology for a catalytic site mutant (E367A) by azide [116]. In the case of *A. saitoi* 1,2-α-D-mannosidase an external nucleophile did not reactivate the D269N and E411A mutant enzymes (Figure 29). Neither Asp269 nor Glu411 residues can be a base catalyst in the normal sense. Although the D269N and E411Q mutant enzymes were catalytically inactive, the D269E and E411D mutant enzymes showed considerable activity. The results indicate that the carboxyl groups at these positions may be required for the 1,2-α-D-mannosidase activity. Structural and modeling studies on cellobiohydrolase Ce16A (EC 3.2.1.91) from *Trichoderma reesei* suggest that the catalytic mechanism may not directly involve a catalytic base [117]. The catalytic mechanism of 1,2-α-D-mannosidase deviates from the typical glycosyl hydrolase.

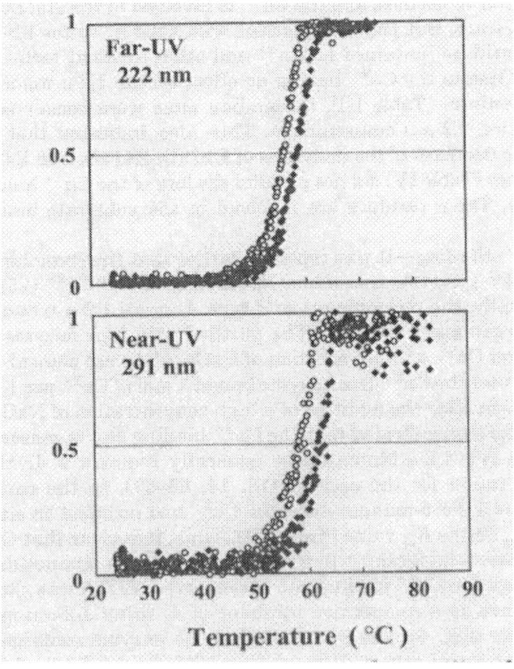

Figure 31. Thermal transition curves of Ca-free (○) and Ca-bound *A. saitoi* 1,2-α-D-mannosidase (◆).

(Tatara, Y. *et al., J. Biol. Chem., 278*, 25289-25294 (2003))

E273D, E414D, and E474D mutant enzymes decreased k_{cat} values and increased K_m values (Table 17). The reduction of the k_{cat} value is assumed to be due to the destabilization of the sugar ring. It has become increasingly clear that ring distortion at the −1 site is crucial in the catalytic mechanism of cellobiohydrolase [118, 119]. The catalytic reaction of α-amylase from *Pseudomonas stuzeri* involves three acidic residues; two residues act as a proton donor and a proton acceptor. The other residue works to tightly bind the substrate, giving a twisted and deformed conformation of the glucose ring at position −1 [120]. Glu273, Glu414, and Glu474 of *A. saitoi* 1,2-α-D-mannosidase may bind the substrate to change the conformation of the mannose ring into more reactive one. The three-dimensional structure of *P. citrinum* 1,2-α-D-mannosidase indicates that Glu271, Glu412, and Glu472 residues (Glu273, Glu414, an Glu474 in *A. saitoi* enzyme) in the active site are located at the bottom of the cavity and are not directly involved in the catalytic mechanism. It is demonstrated that these residues are hydrogen-bound to the Ca^{2+} *via* water molecule, and the Ca^{2+} is involved in the stability of a substrate. However, the recombinant wild type *A. saitoi* 1,2-α-D-mannosidase contained no Ca^{2+} or other divalent metal cations. Despite the Ca^{2+} having no effect on the 1,2-α-D-mannosidase activity (Table 18), its binding sites were conserved in *A. saitoi* 1,2-α-D-mannosidase. This also indicates that the drastic decrease of the activities of E273D, E414D, and E474D mutants (Table 19) did not parallel the loss of the Ca^{2+} binding ability. These residues are involved in the substrate binding sites.

It was reported earlier that the recombinant *A. saitoi* 1,2-α-D-mannosidase has a molecule of Ca^{2+} [112]. In this study [114], the recombinant wild type *A. saitoi* 1,2-α-D-mannosidase contained no Ca^{2+}. The purified wild type enzyme did not bind Ca^{2+} with an addition of $CaCl_2$. We found that the Ca^{2+}-free enzyme bound 1 mol of Ca^{2+} per 1 mol of protein after the addition of a high concentration of NaCl. It was also demonstrated that the Ca^{2+} binding site is conserved (Table 19). 1,2-α-D-Mannosidase generally requires a divalent metal cation for the activity [121-125]. In the case of *A. saitoi* 1,2-α-D-mannosidase, the Ca^{2+} had no effect on either the k_{cat} or K_m value (Table 18). Thus, it is clear that Ca^{2+} is not essential for the activity of *A. saitoi* 1,2-α-D-mannosidase, although the Ca^{2+} binding site is conserved. EDTA was shown to behave as a competitive inhibitor of *A. saitoi* 1,2-α-D-mannosidase (Figure 30). This indicates that the enzyme contains no divalent metal cation; 1-Deoxymannojirimycin inhibits 1,2-α-D-mannosidase as a substrate analog. However, EDTA is not an analog. Tris is a competitive inhibitor for 1,2-α-D-mannosidase from rabbit liver [122]. Tris and other buffers contamining primary hydroxyl group substantially decreased its activity [126, 127]. Carboxyl groups of EDTA may have an affinity for the addition of *A. saitoi* 1,2-α-D-mannosidase. A molecule of Ca^{2+} contributed to thermal stabiliy (Figure 31). The role of Ca^{2+} for *A. saitoi* 1,2-α-D-mannosidase is to protect the enzyme against thermal denaturation, and it was observed to have the same role in *Saccharomyces cerevisiae*-processing 1,2-α-D-mannosidase [128]. Although Ca^{2+} is essential for *S. cerevisiae* enzyme activity [125, 128], *A. saitoi* 1,2-α-D-mannosidase did not require Ca^{2+} for the activity (Table 18). This is the first report demonstrating that 1,2-α-D-mannosidase requires no divalent metal cation for its activity [114].

Table 19. Ca^{2+} binding properties of wild type and mutant *A. saitoi* 1,2-α-D-mannosidases after Ca^{2+}-treatment determined by atomic absorption spectrophotometry

Mutant 1,2-α-D-mannosidase	Bound Ca^{2+} (mol/mol)
Wild type	0.9
E124Q	0.44
E124D	0.8
D269N	0.12
D269E	0.2
D273Q	0
E273D	0
E411Q	0.06
E411D	1.1
E414Q	0
E414D	0
E474Q	0.45
E474D	0.5
E504Q	0.9

(Tatara, Y. *et al., J. Biol. Chem., 278*, 25289-25294 (2003))

A.saitoi 1,2-α-D-mannosidase contains three conserved cysteine residues (Cys334,Cys364, and Cys443). We showed that Cys334 and Cys363 are involved in a disulfide bond and that Cys443 contains a free thiol group [129].

7.4. A Single Cysteine Residue and Disulfide Bond of 1,2-α-D-Mannosidase

For identification of a cysteine residue containing a free thiol group, trypsin-digested peptides of *A. saitoi* 1,2-α-D-mannosidase were reacted with 4-fluoro-7-sulfamoyl-benzofurazan (ABD-F) and separated by reverse-phase HPLC. A single peak was detected based on fluorescence intensity and was analyzed by N-terminal sequencing, and identified as ^{430}DWIWSAFSAV^{439}. The trypsin-digested peptide labeled by ABD-F contained Cys443. The result indicates that Cyss443 contains a free thiol group and that Cys334 and Cys363 are involved in a disulfide bond.

Kinetic characterization of wild-type enzyme (WT), C334A, C363A, and C443A was performed at pH 5.0 using Man-α1,2Man-OMe as a substrate. The results are shown in Table 20. Although the K_m and k_{cat} values of C334A were lower than the WT values, no significant differences in catalytic efficiency (k_{cat}/K_m) were observed between the WT and its variants: C334A, C363A, and C443A. Hence, it was concluded that the mutations at three cysteine residues did not affect the catalytic performance of 1,2-α-D-mannosidase. Hence, it was concluded that the mutations at Cys334, Cys363, and Cys443 did not affect the catalytic performance of *A. saitoi* 1,2-α-mannosidase.

For thermostability of cysteine mutants, wild-type (WT), C334A, C363A, and C443A were incubated at 55°C, and the half-lives of residual activity ($t_{1/2}$) at 30°C were 119, 5.2, 11.3, and 24.4 min respectively (Table 20). Mutation at Cys334, Cys363, and Cys443 drastically decreased the thermostability of 1,2-α-D-mannosidase. It is evident that the thermostability of *A. saitoi* 1,2-α-D-mannosidase is significantly influenced by mutations.

The wild-type (WT) and cysteine mutants bound about one mole of Ca^{2+} per mole of protein after Ca^{2+}-binding treatment (Table 20). The Ca^{2+}-bound forms were also examined for thermostability, as shown in Figure 32. The half lives of the Ca^{2+}-bound enzymes increased slightly, as compared to those of the Ca^{2+}-free enzyme.

Table 20. Kinetic parameters, Ca^{2+}-binding properties, and thermostabilities of wild-type (WT) and cysteine mutants of *A. saitoi* 1,2-α-mannnosidase

Enzyme	K_m	k_{cat}	k_{cat}/K_m	Ca^{2+}-binding[a]	$t_{1/2}$[b]	T_m[c]
	(mM)	(s-1)	($s^{-1} \cdot M^{1}$)	(mol/mol)	(min)	(°C)
Wild type	0.719 ± 0.013	8.51 ± 0.09	11.8	0.0	119	55.8
				1.1	125	
C334A	0.296 ± 0.006	3.87 ± 0.12	13.0	0.0	5.2	49.6
					6.8	
C363A	0.780 ± 0.007	9.48 ± 0.03	12.3	0.0	11.3	49.6
					18.8	
C443A	0.719 ± 0.007	6.67 ± 0.04	9.3	0.0	24.4	51.9
				1.2	38.3	

[a] *A. saitoi* 1,2-α-mannnosidase binds one mole of Ca^{2+} per mole of protein after Ca^{2+}-binding treatment [129].

[b] $t_{1/2}$ is the half-life time (min) when the residual activity of 1,2-α-mannnosidase is 50% after incubation at 55°C.

[c] T_m is a transition midpoint (°C), which is an apparent value because the thermal denaturation is not reversible.

(Tatara, Y. *et al., Biosci. Biotechnol. Biochem., 69*, 2101-2108 (2005))

Ca^{2+} is essential for the activity of 1,2-α-D-mannosidase family members, such as *S. cerevisiae* [128] and human endoplasmic reticulum [121], but the *A. saitoi* 1,2-α-D-mannosidase is not dependent on Ca^{2+} and has no divalent metal cation [114]. Previously we found that it bound one mol of Ca^{2+} per mol of protein at the active center, which contributed to thermostability [112]. The cysteine mutants also bound Ca^{2+} and were stabilized to some extent (Table 20, Figure 32), though the binding of Ca^{2+} did not fully recover the drastic decrease in thermostability caused by the cysteine mutations [129].

Twelve kinds of mutations at position 443 replacing the Cys443 with alanine, asparagine, aspartic acid, glycine, isoleucine, leucine, methinine, phenylalanine, serine, threonine, tyrosine, and valine were constructed in the *msdS* gene and expressed by *A. oryzae*. Seven mutant enzymes, C443A, C443D, C443G, C443L, C443S, C443T, and C443V, were purified and assayed for 1,2-α-D-mannosidase activity at pH 5.0 using Man-α1,2Man-OMe as a substrate. The specific activity of each mutant enzyme is shown in Table 21.

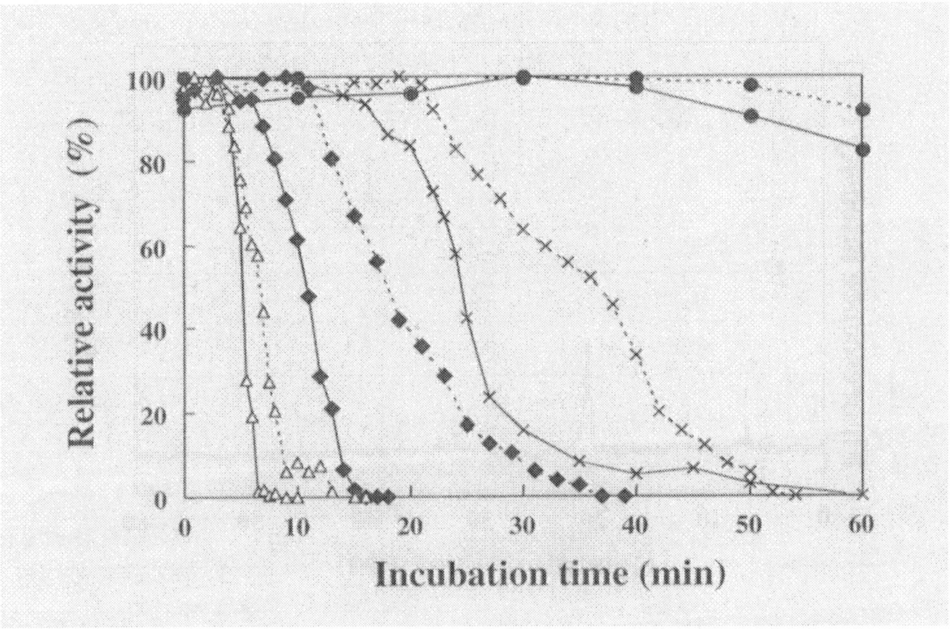

Figure 32. Thermostability of wild type and cysteine mutants of *A. saitoi* 1,2-α-D-mannosidase at 55°C, and the effect of Ca^{2+}-binding. The enzyme solution was incubated at 55°C, and the residual activity was assayed. The data shown correspond to WT (●), C334A (△), C363A (◆), and C443A (×). Lines (—) and dotted lines (┅) correspond to Ca^{2+}-free and Ca^{2+}-bound enzymes respectively.

(Tatara, Y. *et al., Biosci. Biotechnol. Biochem., 69*, 2101-2108 (2005))

The specific activities of C443A, C443G, C443S, and C443T were almost the equal of that of the WT. The relative activities of C443D, C443L, C443M, and C443V were less than 50%, as compared to the WT. The other four mutant enzymes, C443F, C443I, C443N, and C443Y, were undetectable or slightly detectable by Western blot analysis of the culture medium.

Far-UV and near-UV spectra were measured to analyze the secondary and tertiary structures respectively. The polypeptide backbone is optically active in the far-UV (250-170 nm), and different secondary structures produce characteristic spectra. Tryptophan, tyrosine, and phenylalanine can give CD signals in the near-UV (300-270 nm). The constructed free rotation of the side chains exhibits the absorption band.

Table 21. Specific activities and melting temperatures (T_m) of wilt type (WT) and Cys443 variants of *A. saitoi* 1,2-α-mannnosidase, and comparison of amino acid properties

Amino acids at position 443	Specific activity (m katal/kg)	T_m[a] (°C)	Van der Waals volume[b] (Å^3)	Hydro-phobicity[c]
Cys (WT)	47.4	55.8	86	+2.5
Thr	44.9	52.8	93	-0.7
Ala	38.3	51.9	67	+1.8
Gly	44.0	50.2	48	-0.4
Ser	47.1	50.0	73	-0.8
Asp	19.8	ND[d]	91	-3.5
Val	20.7	ND	105	+4.2
Met	14.5	ND	124	+1.9
Leu	11.7	ND	124	+3.8
Asn	NA[e]	NA	96	-3.5
Ile	NA	NA	124	+4.5
Phe	NA	NA	135	+2.8
Tyr	NA	NA	141	-1.3

[a]The midpoint of the transition is an apparent value because thermal denaturation is not reversible.

[b]The van der Waals volumes are from Richars, F. N., *J. Mol. Biol., 82*, 1-14(1974).

[c]Hydrophobicity indices are from Kyte, J. & Doolittle, R. F. A., *J. Mol. Biol., 157*,105-132(1982).

[d]ND means that the values are not determined, because the mutant enzymes from denatured structure.

[e]NA means that the mutant enzymes were not analyzed, because they were not secreted in the culture medium.

(Tatara, Y. *et al., Biosci. Biotechnol. Biochem., 69*, 2101-2108 (2005)).

The spectra revealed that WT, C334A, and C363A had the same minima and maxima, indicating that they had similar secondary and tertiary structures.

Consequently, the disulfide bond (C334 and C363) does not contribute significantly to the structure of the enzyme. The far-UV and near-UV spectra of the WT and Cys443 variants (C443A, C443D, C443G, C443L, C443S, C443T, and C443V) were determined. The spectra of the WT, C443A, C443G, C443S, and C443T presented the same waveform and mean residue ellipticity, indicating similar secondary and tertiary structures. Meanwhile, large decreases in the CD intensity of C443D, C443L, C443M, and C443V were observed over the whole range of wavelengths. These results might be ascribed to the partial unfolding of the mutant enzyme.

The CD signal at 222 nm as a function of temperature was collected and fitted to two-state models for WT, C334A, and C363A (Figure 33). The parameters from those fits were used to calculate a curve representing the apparent fraction of unfolded enzyme, F_{app}. The apparent denaturation midpoint (T_m) was determined. Both C334A and C363A exhibited T_m of 49.6°C, 6.2°C lower than that of the WT. The thermal unfolding curves of the Cys443 variants were determined. As compared to the wild type 1,2-α-D-mannosidase (WT), the T_m values decreased by 3.9, 5.6, 5.8, and 30°C for C443A, C443G, and C443S, respectively, (Table 21). The C443T mutant retained the most stability. The thermostability of C443D, C443L, C443M, and C443V was not analyzed because these mutants would have very little tertiary and secondary conformation.

We demonstrated with *Aspergillus* 1,2-α-D-mannosidase that Cys334 and Cys363 forms a disulfide bond and that Cys443 contains a free thiol group. These results are consistent with the crystallographic data on *Penicillium* 1,2-α-D-mannosidase in that the corresponding cysteine residues are involved in the formation of disulfide bond [115]. The structure of *A. saitoi* 1,2-α-D-mannosidase was constructed and viewed with a Swiss-Pdb Viewer. Figure 34 shows the Cys443 and its surrounding amino acid residues. The hydrophobic side chains of Ile375, Leu408, Ala438, Val439, Phe449, and Leu452 are present within 6 Å of the S^{γ} atom of C443. The free thiol group of Cys443 is not involved in a hydrogen bond to any amino acid residues.

Twelve mutant enzymes of Cys443 that contained a large side chain at position 443, such as C443F, C443I, and C443Y, were not secreted into the culture medium. The van der Waals volumes of phenylalanine, isoleucine, tyrosine, and cysteine are 135, 124, 141, and 86 $Å^3$ respectively (Table 21). It is suggested that large amino acids are not preferred at position 443 because they cause a failure of protein folding *in vivo*. In eukaryotic cells, misfolded proteins are selectively degraded to prevent the accumulation of non-functional, potentially toxic proteins [130-132]. Mutant enzymes such as

C443F, C443I, and C443Y expressed in *A. oryzae* are recognized as misfolded proteins by the quality control system in the endoplasmic reticulum. The C443N mutant was not secreted into the culture medium either, even though the van der Waals volume of asparagine is comparatively small (96 Å). The introduction of asparagine at position 443 might give rise to a potential asparagine-linked glycosylation site (Asn443-Arg444-Thr445). The asparagine residue at position 443 was possibly glycosylated. Oligosaccharides linked to asparagine residues in the context of Asn-Xaa-Ser/Thr are known to have essential roles as carriers of biological information [133]. The unfavorable asparagine-linked sugar chain probably caused the disturbance of protein folding of secretion.

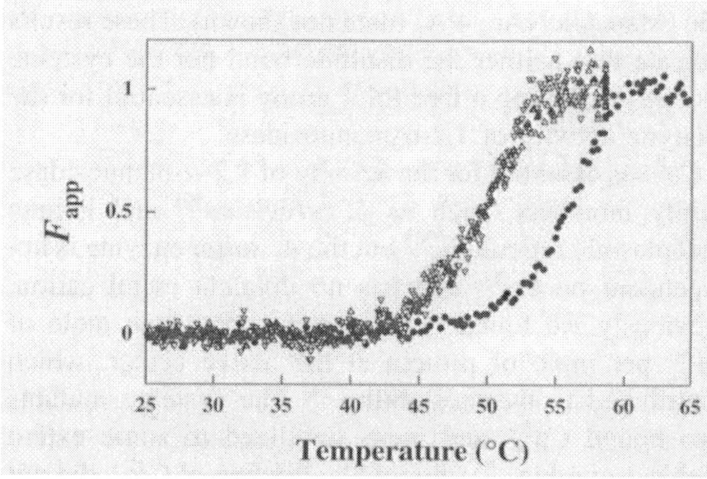

Figure 33. Thermal unfolding curves of WT (●) and cysteine mutants, C334A (▽) and C363A (△), of *A. saitoi* 12,-α-D-mannosidase.

(Tatara, Y. *et al., Biosci. Biotechnol. Biochem., 69*, 2101-2108 (2003))

The specific activity of C443D, C443L, C443M, and C443V was less than 50%, compared to that of the WT (Table 21). The CD spectra of these mutants indicated that they were partly unfolded. The van der Waals volumes of aspartic acid (91 Å^3) and valine (105 Å^3) are close to that of cysteine (86 Å^3). Aspartic acid and valine have bifurcated side chains at the C^{γ} and C^{β} atoms respectively. The bifurcated side chains are assumed to cause denaturation of the enzyme.

Leucine and isoleucine have the same van der Waals volume (124 Å^3), and have bifurcated side chains at the C^{γ} and C^{β} atoms respectively. The C443L mutant was secreted as demonstrated structure, while C443I was not secreted. These results suggest that the bifurcated side chain at C^{β} of isoleucine has a large side steric hindrance for the current protein folding and causes the degradation of C443I. Methinine is a sulfur-containing amino acid like cysteine. The C443M mutant also had a partly unfolded structure, which was thought to be due to the large van der Waals volume of methioine (124 $Å^3$).

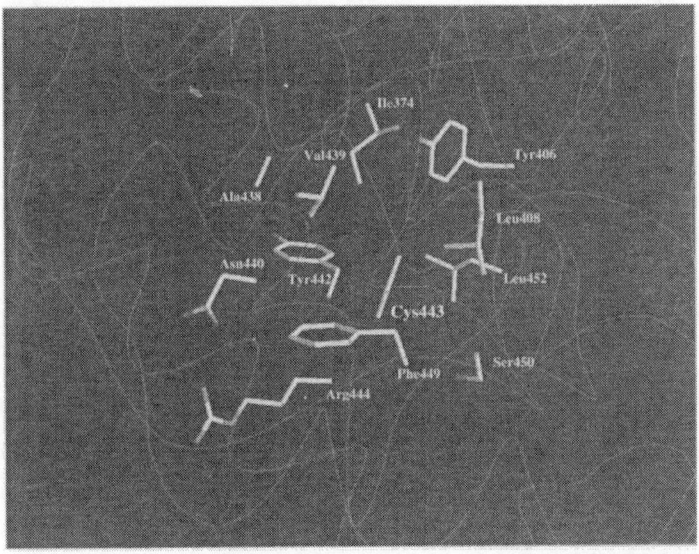

Figure 34. Structural model of *A.saitoi* 1,2-α-D-mannosidase focused on the side chain of C443 and its surrounding amino acids. The structural model was obtained using the SWISS-MODEL server. The side chains within 6 Å of the S^{γ} atom of the Cys443 are shown as a stick representation. The main chain is also represented as a line.

(Tatara, Y. *et al., Biosci. Biotechnol. Biochem., 69*, 2101-2108 (2003))

When the cysteine443 was replaced with Ala, Gly, Ser, or Thr, the mutant enzymes folded the current conformations. Cys443T was the most thermostable mutant enzyme (Table 21). In this case the van der Waals volume of threonine is 93 $Å^3$, slightly larger than that of cysteine (86 $Å^3$). On the other hand, the hydrophobicities of cysteine and threonine are +2.5 and -0.7 respectively [134]. The decrease in the T_m value of the C443T mutant is thought to be due to the low hydrophobicity of threonine.

The C443A mutant is the second most thermostable mutant enzyme. Cysteine and alanine are hydrophobic amino acids (the hydrophobicity of alanine is +1.8). C443A was more thermostable than C443S, although the van der Waals volume of serine (73 $Å^3$) was closer to that of cysteine (86 $Å^3$) than alanine (67 Å^3). These results indicate that hydrophobicity at position 443 is crucial to the thermostability of the enzyme. The C443G mutant was also thermolabile. The replacement of cysteine with glycine introduces a void in the space at position 443, which appears to destabilize the enzyme. From these results, we deduced that Cys443 of *A. saitoi* 1,2-α-D-mannosidase is involved in a hydrophobic interaction with the surrounding residues to stabilize the enzyme, and that the van der Waals volume of the amino acid is important to retain a stable interaction.

Molecular modeling of *A. saitoi* 1,2-α-D-mannosidase showed that Cys443 was located in the loop connecting the α13 and α14 helices as well as Cys441 of *Penicillilum* 1,2-α-D-mannosidase. The side chain of Cys443 was buried and surrounded by a hydrophobic environment (Figure 34). When the hydrophobic interaction around C443 was disrupted by site-directed mutagenesis, the normal folding of the enzyme was disturbed. The misfolded and partly denatured mutants were degraded (C443I, C443F, C443Y) and secreted (C443D, C443V, C443M, and C443L) respectively. It is suggested the alteration of the local structure around Cys 443 affects normal folding and determines the destination of the protein.

It is thought that Cys443 is buried to avoid exposure of its highly reactive free thiol group. Further, it is suggested that Cys443 greatly contributes to the thermostability of the enzyme by increasing hydrophobic packing.

7.5 Production of Human Compatible High Mannose-Type ($Man_5GlcNAc_2$) Sugar Chain in Yeast Cells

A yeast mutant capable of producing $Man_5GlcNAc_2$ human compatible sugar chains on glycoproteins was constructed [135]. An expression vector for 1,2-α-D-mannosidase with the "HDEL" endoplasmic reticulum retention/retrieval tag was designed and expressed in *Saccharomyces cerevisiae*. An *in vitro* 1,2-α-D-mannosidase assay and Western blot analysis showed that it was successfully localized in the endoplasmic reticulum. A triple mutant yeast lacking three glycosyltransferase activities was then transformed with an 1,2-α-D-mannosidase expression vector. The oligosaccharides' structure of carboxypeptidase Y as well as surface glycoproteins were analyzed, and the recombinant yeast was shown to produce a series of high mannose-type sugar chains including $Man_5GlcNAc_2$. This is the first report of a recombinant *S. cerevisiae* able to produce $Man_5GlcNAc_2$-oligosaccharides, the intermediate for hybrid-type and complex-type sugar chains.

Chapter 8

ACID ACTIVATION OF PROTYROSINASE FROM *ASPERGILLUS ORYZAE*

8.1. ACID ACTIVATION OF PROTYROSINASE FROM *A. ORYZAE*

Tyrosinase (monophenol monooxygenase or monophenol dihydroxy-L-phenyalanine: oxygen oxidoreductase, EC 1.14.18.1) is the only known enzyme participating in the process of melanin biosynthesis [136]. The first two steps in the pathway are the oxygenation of monophenols to *o*-diphenols and the oxidation of diphenols to *o*-quinones (catechol oxidase or 1,2-benzenediol : oxygen oxidoreductase, EC 1.10.3.1), both of which require molecular oxygen. The copper-containing monooxygenase catalyzes both the *o*-hydroxylation of monophenols to *o*-diphenols and the oxidation of *o*-diphenols to *o*-quinones.

L-tyrosine + dihydroxy- L-phenylalanine (L-DOPA) + O_2 =
L-DOPA + DOPA-quinone + H_2O (EC 1.14.18.1)
2 catechole + O_2 = 2 1,2-benzoquinone + 2 H_2O (EC 1.10.3.1)

These steps are followed by a series of non-enzymatic steps resulting in the formation of melanin [137, 138]. Melanin has many biological functions [139]. Reactive quinone intermediates in the melanin biosynthetic pathway exhibit antibiotic properties. The melanin polymer is an important strengthening element of plant cell walls and insect cuticles. Light absorption by melanin has several biological functions, including photoreceptor shielding, thermoregulation, photoprotection, camouflage and display. Melanin is a powerful cation chelator and may act as a free radical sink. In fungi, melanin plays a role in the reproductive organ, differentiation, spore formation, pathogenic fungi virulence, and tissue protection after injury [140, 141].

Copper ions in the (I) and (II) oxidation states are biologically important. Basically, three different types of copper center are shown. 'Blue' of type I copper occurs in the blue electron carrying proteins such as stellacyanin, plastocyanin and azurin. There is also 'non-blue' or type II copper and a type III copper center that is 'non-detectable' by electron paramagnetic resonance (EPR), apparently containing a pair of contiguous copper atoms. Tyrosinase is a type III protein that is not detectable by EPR, because it is strongly antiferromagnetically coupled in pairs [142]. Type II and III copper forms a trinuclear center to reduce dioxygen to two molecules.

Haneda *et al.* [143, 144] found that, when intact cells or an enzyme solution of *A. oryzae* IAM 2625 were used, L-DOPA synthesis from L-tyrosine could proceed preferentially under acidic conditions, around pH 3—4. Since L-DOPA attracted attention as an effective drug for Perkinson disease, reports concerning its microbial production have been published [145]. Further studies [146] led to the presumption that the enzyme of *A. oryzae* having L-DOPA-forming activity might exist *in vivo* as an inactive form, since the formation of L-DOPA was markedly activated by acidifying treatment below pH 5. The only reports on the phenomenon of the activation of tyrosinases and other oxygenases by acidifying treatment are found in references [144, 146].

Our previous report [147] described the unique process of the activation of intracellular protyrosinase isolated from *A. oryzae* RIB 128 (designated as ATCC 22788) by acidifying treatment at pH 3.0. *A. oryzae* was grown in 500-ml Sakaguchi flasks containing 100 ml modified Czapeck medium. Three forms of protyrosinase—I, II, and III—were isolated by disc-gel electrophoresis at pH 9.4. Purified protyosinase I and III each migrated as single band on disc-gel electrophoresis at pH 9.4, although the band of tyrosinase stained with L-DOPA after acidifying treatment was broader. When fresh mycelia were used, about 42 % of the total protyrosinase were located in the protyrosinase I fraction. Experimental evidence showed that the purified protyrosinase is contaminated with firmly attached proteinases, aspartic proteinase and serine proteinase and that contamination may account for both the multiple molecular forms of protyrosinase/tyrosinase and the activation of the proenzyme at acidic pH.

The three forms of protyrosinase exhibited similar molecular masses, estimated to be 240,000 Da by gel filtration with Sephadex G-200 [147].

Interestingly, activated tyrosinase was absorbed at the top of the dextran gel of a Sephadex G-200 column but protyrosinase was not. The molecular mass of tyrosinase I, II, and III was estimated to be 150,000 Da by gel filtration on Toyopearl HW-55 [147]. The circular dichroism (CD) signal of the activated tyrosinase was greatly changed compared with that of protyrosinase in the range of 203-250 nm, indicating that the conformation of protyrosinase was changed after acid activation [147] These results assumed that protyrosinase and tyrosinase have different conformation, different stabilities and some different properties.

The activation pH of protyrosinase was in the range of pH 2.0-4.0, and the optimal pH of the activation was 2.5-3.0; little or no activation was detected above pH 5.0 and below pH 1.0. With this treatment, about 80% of total protyrosinase could be converted into activated tyrosinase within 2-3 min; thus, the activation reaction was very rapid. A Dixon plot of the acid-activated tyrosinase (acid-activated tyrosinase I) with mimosine as an inhibitor showed competitive inhibition. The inhibitor constant, *K*i, for mimosine was 0.13 μM. Competitive inhibitions were observed with tropolone, kojic acid, and benzoic acid, and the *K*i values for which were 0.22 μM, 12 μM, and 700 μM, respectively [148].

The coding region of the protyrosinase gene, *melO*, from *A. oryzae* occupies 1671 base pairs (bp) of the genomic DNA and is separated into two exons by one intron [149]. The full-length cDNA of the *melO* gene was cloned. Analysis of the 1671-bp nucleotide sequence revealed a single open reading frame encoding 539 amino acid residues. The cDNA has been expressed in yeast *Saccharomyces cerevisiae* cells by using the expression plasmid pG3TY. The predicted protein product derived from the *melO* gene is identified by Western blotting and activity determination.

The predicted amino acid sequence of the gene product was compared with that of *Neurospora crassa* tyrosinase [150]. A coupled pair of three histidine residues (His-63, His-84 and His-93 for Cu(A) and His-290, His-294 and His-333 for Cu(B)) of protyrosinase from *A. oryzae* was assumed to correspond to Cu(II) ligands in the homologous tyrosinases from *Streptomyces glaucescens* [151, 152] and *Homo sapiens* [153]. The predicted amino acid sequence of the *melO* gene product has 17% and 15% homologous with those from *N. crassa* [154] and *Homo sapiens* [153], respectively.

The expression product of pG3TY was also isolated by PAGE at pH 9.4 gel [149]. The activity band of tyrosinase was stained with dihydroxy-L-phenylalanine (L-DOPA) after acidifying at pH 3. It can be concluded that the expression product of the *melO* gene is the protyrosinase of *A. oryzae*.

8.2. Homo-Tetrameric Protyrosinase is Converted to Active Dimers with an Essential Intersubunit Disulfide Bond at Acidic Ph

The quaternary structure of tyrosinase differs depending on the species. The tyrosinase can be found in either its latent or active form, and the activating conditions differ depending on the species. Tyrosinase from the mushroom *Agaricus bisporus* is a heterotetramer that is a compound of two subunits of 43 kDa (H) and two subunits of 13 kDa (L), with a native molecular mass of 111 kDa, and contains four copper atoms [155, 156]. The isolated subunits do not possess the enzymatic activity. Activation of the mushroom pro-tyrosinase can be effected by protease treatment [157] or by SDS [158].

In the case of the filamentous fungus *Neurospora crassa*, the tyrosinase is thought to be a 46 kDa monomer with full activity that is formed upon proteolytic processing of a 75 kDa precursor [150] *Streptomyces* tyrosinase is a monomer of 31 kDa with enzymatic activity [151, 152]. The tyrosinase from *Homo sapiens* melanoma cell is a monomer of 53 kDa that has enzymatic activity [159]. *Xenopus laevis* tyrosinase is activated by anionic detergents such as SDS [160]. Protyrosinase from the epidermis of the frog *Rana esculenta ridibunda* can be activated *in vitro* by several proteinases (trypsin, alpha-chymotrypsin, Pronase) and by light [161]. Furthermore, the protyrosinase and tyrosinase have different quaternary structures; the dimeric structure of protyrosinase is known to be converted to a tetramer after the activation by trypsin.

Aspergillus oryzae protyrosinase (pro-TY) is a homotetramer composed of 60 kDa subunits containing 2 mol of copper per mol of subunit [149]. The proenzyme is inactive at neutral pH, and is activated at acidic pH [147]. The acid activation is quite a unique mechanism for the activation of a protyrosinase. To investigate the mechanism of acid activation, and the conformational change following the change of pH and interactions between the subunits were examined [162].

The acid activation of protyrosinase from *A. oryzae* was the most potent at pH 3.0, and the activity was increased logarithmically with increasing acid activation time, with a rate constant of 1.5 x 10^{-3}/s. An anionic detergent (0.01% SDS) slightly activated protyrosinase. The specific activity of the SDS-activated tyrosinase was 19% compared with that of acid-activated tyrosinase (acid-TY). Protyrosinase was not activated by the other reagents tested, including Tween 20 (0.001-0.1%), urea (0.5-2.0 M), NaCl (0.1-2.0 M), acetonitrils (5-20%), ethanol (5-20%), and protease (trypsin and chymo-trypsin).

It is thought that polar interactions are disrupted by protonation at acidic pH or by an anionic detergent, e.g., SDS. Ionic bonds and/or hydrogen bonds have been shown to be involved in the interaction between the dimers. A non-ionic detergent, Tween 20, did not activate protyrosinase, indicating that the contribution of hydrophobic interactions in the dimer-dimer interface was minor. The possibility of processing by a contaminated acid protease was ruled out by SDS-PAGE analysis of acid-activated tyrosinase is not shift to a lower molecule. It is suggested that protyrosinase is not cleaved for activation by acid treatment. Atomic absorption spectroscopy showed that protyrosinase and acid-activated tyrosinase contained 2.0 and 0 mol of copper per mol of subunit, respectively, after the treatment with ethylenediaminetetraacetic acid (EDTA). These results indicate that a conformational change causes exposure of the binuclear active copper center.

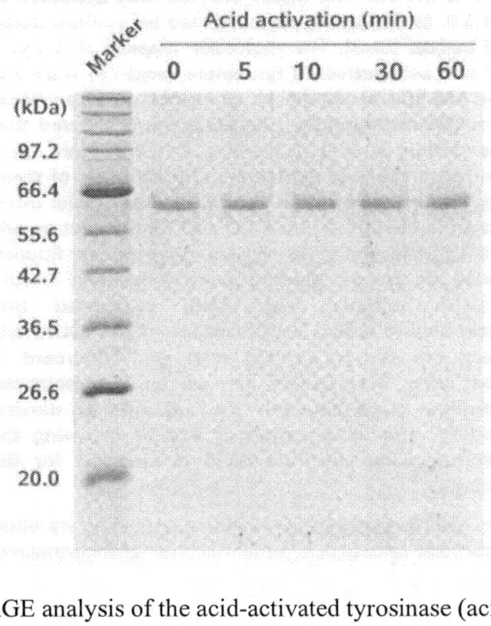

Figure 35. SDS-PAGE analysis of the acid-activated tyrosinase (acid-TY) showing the time course of acid activation at pH 3.0.

(Tatara, Y. *et al.*, *Pigment Cell Melanoma Res.*, *21*, 89-96 (2008))

The relative molecular mass of both protyrosinase and acid-activated tyrosinase was 60 kDa, as shown by SDS-PAGE (Figure 35). This result was in agreement with the calculated molecular mass of protyrosinase (60,604 Da). Protyrosinase and acid-activated tyrosinase did not migrate in the polyacrylamide gel in the absence of the reducing agent because of thiol-disulfide interchange reactions that occur between the SDS-denatured proteins.

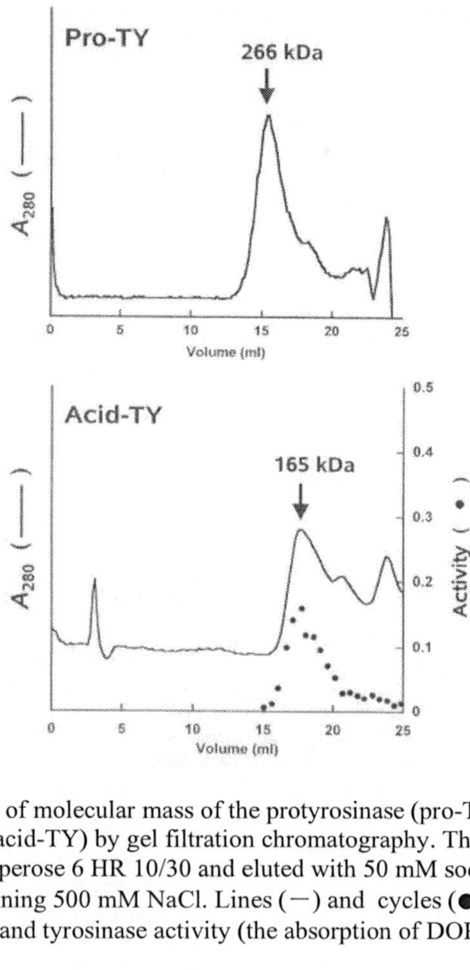

Figure 36. Estimation of molecular mass of the protyrosinase (pro-TY) and acid-activated tyrosinase (acid-TY) by gel filtration chromatography. The pro-TY and acid-TY was applied to Superose 6 HR 10/30 and eluted with 50 mM sodium phosphate buffer (pH 7.0) containing 500 mM NaCl. Lines (—) and cycles (●) represented the absorption at 280 nm and tyrosinase activity (the absorption of DOPA chrome at 475 nm), respectively.

(Tatara, *et al., Pigment Cell Melanoma Res., 21*, 89-96 (2008))

The native molecular masses of protyrosinase and acid-activated tyrosinase were 266 and 165 kDa, respectively, as estimated by gel-filtration chromatography (Figure 36, Table 22) [162]. Protyrosinase was judged to be a tetramer with a calculated molecular mass of 242,416 Da. It was assumed that the quaternary structure of the acid-activated tyrosinase is dimeric.

Table 22. Specific activity and molecular mass under various conditions of protyrosinase and acid-activated tyrosinase from *A. oryzae*

Condition	Molecular mass (kDa)	Quaternary structure	Specific activity (katal/kg)	Specific activity (%)
Protyrosinase	266	Tetramer	ND	—
Tyrosinase	165	Dimer	1.2	100
Reduced	68[a]	Monomer	—	—
Oligo-C108A	380	Oligomer	0.016	1.3
Mono-C108A	68	Monomer	ND	—

ND, not detected.

[a]Gelfiltration chromatography under reducing conditions was performed in the presence of 2% β-mercaptoethanol at pH 7.0.

(Tatara, Y. *et al., Pigment Cell Melanoma Res., 21*, 89-96 (2008))

The apparent molecular mass of acid-activated tyrosinase was increased because of partial unfolding by the acidic pH. The molecular mass of protyrosinase under reducing conditions was 68 kDa, indicating that the quaternary structure was monomeric. Our data suggest that the intersubunit polar interaction was disrupted at pH 3.0, and the tetrameric protyrosinase dissociated to dimers. The binuclear copper of the active center was then exposed.

The CD spectra in the far-UV region revealed that pro-tyrosinase and acid-activated tyrosinase had similar secondary structures (Figure 37). On the other hand, the CD signal of acid-activated tyrosinase was greatly decreased compared with that of protyrosinase in the range of 280-290 nm, indicating that the tertiary and/or quaternary structure of protyrosinase was changed by acid treatment. On the basis of these results, we deduced that the intersubunit polar interaction is disrupted at pH 3.0, and that the tetrameric protyrosinase dissociates to dimmers [162].

To detect possible conformational changes caused by acid activation, tryptophan fluorescence spectra were measured. When excited at 295 nm, the emission spectrum of protyrosinase exhibited an average emission wavelength of 339 nm (Figure 38). By comparison, in the spectrum of acid-activated tyrosinase, the emission intensity was decreased by 62%, and there was a red shift of the average emission wavelength to 346 nm. These results indicate that tryptophan residues are brought into a polar environment by acid activation.

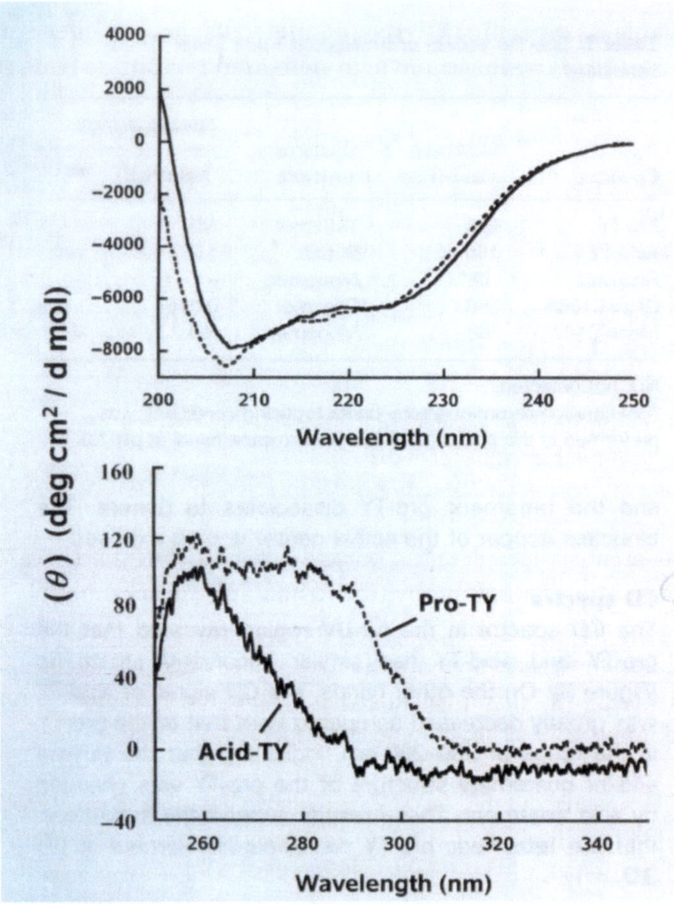

Figure 37. Comparison of CD spectra of the pro-tyrosinase (pro-TY, ---) and acid activated-tyrosinase (acid-TY, —).

(Tatara, Y. *et al., Pigment Cell Melanoma Res., 21*, 89-96 (2003))

8-Anilino-1-naphathalene sulfonic acid (ANS) is a sensitive hydrophobic patch probe [163]. Protyrosinase did not bind ANS, whereas the acid-activated tyrosinase did bind ANS as indicated by an increase in the fluorescence intensity (Figure 39). These results reveal that the acid activation exposes the hydrophobic surface to the solvent. The dissociation of tetrameric protyrosinase is followed by the exposure of the binuclear active copper center and hydrophobic amino acids.

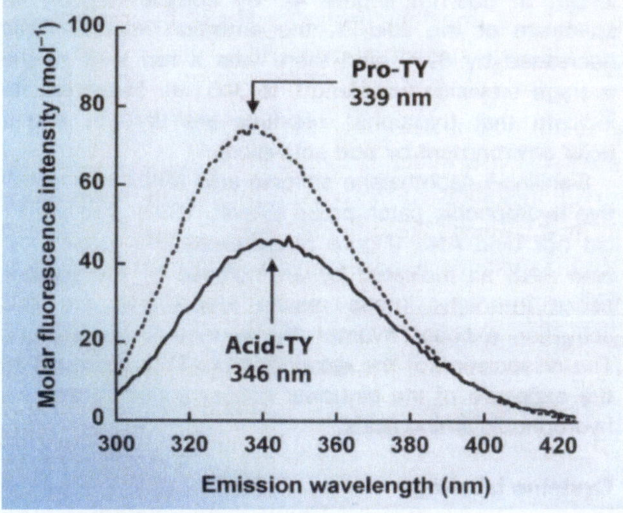

Figure 38. Tryptophan fluorescence spectra of the protyrosinase (pro-TY) (---) and acid- activated tyrosinase (acid-TY) (—).

(Tatara, Y. *et al., Pigment Cell Melanoma Res., 21*, 89-96 (2008))

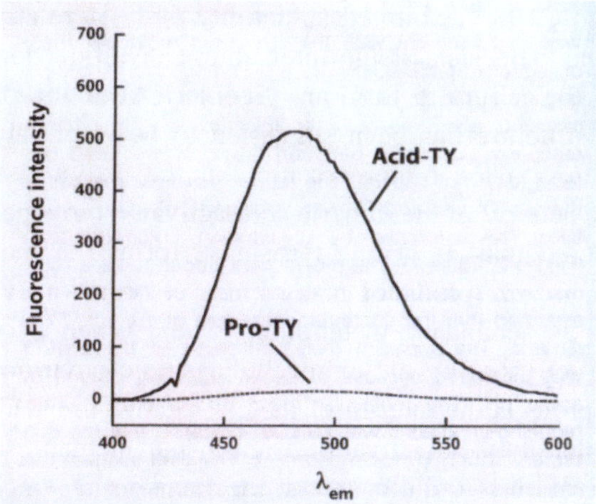

Figure 39. 8-Anilino-1-naphthalene sulfonic acid (ANS) binding assay of the protyrosinase (pro-TY, ---) and acid-activated tyrosinase (acid TY, —).

(Tatara, Y. *et al., Pigment Cell Melanoma Res., 21*, 89-96 (2008))

Protyrosinase from *A. oryzae* has eight cysteine residues [149]. Cys-82 is proposed to be involved in the ligation of CuA in the active center or to be covalently linked via a thioether bridge to His84 by analogy to *Neurospora crassa* [154] and the crystal structure of the sweet potato catechol oxidase [164]. At least one cysteine contains a free thiol group or forms an intermolecular disulfide bond in protyrosinase from *A. oryzae*. The quaternary structure of protyrosinase under reducing conditions was monomeric (Table 22). This observation suggested the possibility that an intermolecular disulfide bond can be formed between the subunits.

To identify the cysteine residues involved in the intermolecular disulfide bond, free thiol groups were labeled with a fluorogenic reagent, 4-fluoro-7-sulfamoylbenzofurazan (ABD-F) [165], in the presence of a stepwise addition of a reducing agent. Cys108 probably formed an intermolecular disulfide bond between the subunits of dimeric acid-TY (Figure 40). Dimerization of acid-activated tyrosinase involving the intermolecular disulfide bond is essential for its activity [162].

The purified Cys108Ala (C108A) mutant migrated as a 60 kDa band on the SDS-PAGE gel. The C108A mutant was eluted with gel-filtration chromatography to investigate the quaternary structure (Table 22). The molecular mass of one was 68 kDa and that of the other was 380 kDa. The former was suggested to be a monomer (mono-C108A) and the latter was an oligomer (oligo-C108A). The mono-C108A was enzymatically inactive, whereas the oligo-C108A exhibited an activity of 1.3% relative to that of the wild type enzyme. The lack of the intersubunit disulfide bond is thought to have resulted in the failure to form an oligomeric structure of mono-C108A. The oligomeric structure has been suggested to be essential for its activity. Cys108 contributes to the oligomerization of the enzyme, and hence, it is important for the activity of the dimeric acid-activated tyrosinase.

It is considered that *A. oryzae* tyrosinase forms the inactive tetrameric conformation. *A. oryzae* produces many kinds of organic acids such as kojic acid and citrate as secondary metabolites [166]. When a cell is damaged, protyrosinase is activated by the extracellular acidic environment to produce melanin, which protects the interior of the cell against the acidic environment. It is considered that the acid activation is an instantaneous response to an acid stress. Acid activation is quite a unique mechanism, in which the tetrameric proenzyme is activated by dissociation to dimers resulting from the acidic shift of pH. Acid-activated tyrosinase remains in the dimeric form via an intersubunit disulfide bond, which is probably formed between the Cys108 residues. The dimerization of the acid-activated tyrosinase from *A. oryzae* is essential for the activity of the tyrosinase.

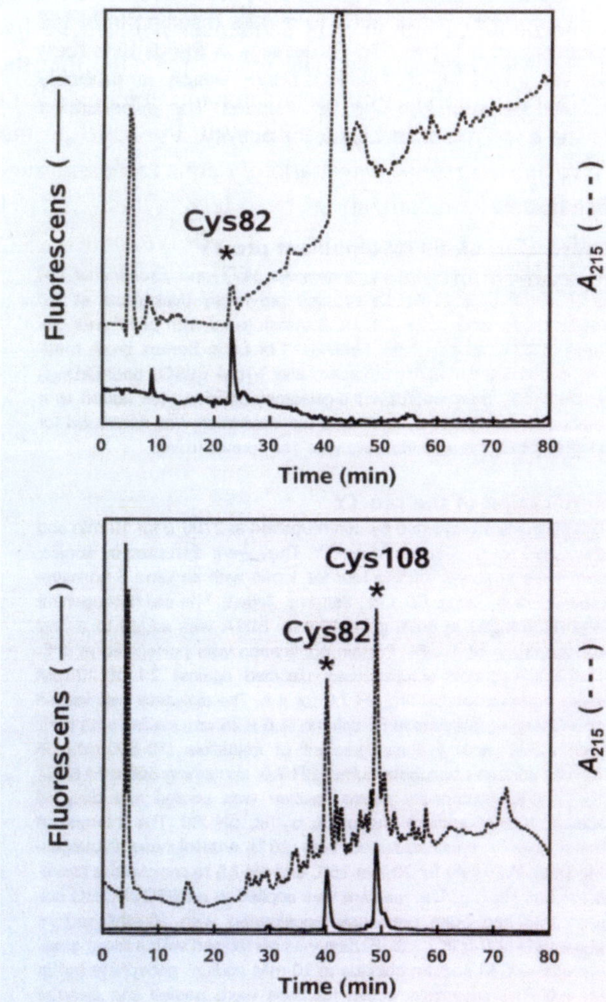

Figure 40. Reverse-phase HPLC of trypsin-degraded protyrosinase (pro-TY) in the absence (0 mM DTT, upper panel)) and presence (1 mM DTT, lower panel) of reducing agent (4-fluoro-7-sulfamoylbenzofurazan, ABD-F).

(Tatara, Y. *et al., Pigment Cell Melanoma Res., 21*, 89-98 (2008))

In this connection, we have cloned a novel tyrosinase-encoding gene (*melB*) specifically expressed in solid-state culture (Koji making) of *A. oryzae* [167]. Molecular cloning showed that the gene consisted of six exons interrupted by five introns and had an open reading frame encoding 616 amino acid residues. The deduced amino acid sequence of the gene had only 24% homology with the MelO tyrosinase. The results revealed that the *melB* gene was highly expressed in solid-cultures but not in submerged cultures. When the *melB* gene was expressed under the control of the *melB* promoter in solid-state cultures, tyrosinase activity was markedly enhanced, and the culture mass was brown because of melanization by MelB tyrosinase. These results indicated that the *melB* gene encodes a novel tyrosinase associated with melanization in solid-cultures of *A. oryzae* [167].

Chapter 9

Hyper Production System of *Aspergillus Oryzae* Glucoamylase in Submerged Culture under Tyrosinase-Encoding Gene (*MelO*) Promoter Control

9.1. Glucoamylase from *A. Oryzae*

Glucoamylase (α-1,4-glucan glucanohydrolase, EC 3.2.1.3) derived from filamentous fungus *A. oryzae* strains is used in manufacturing the Japanese alcoholic beverages saké (rice wine) and shochu (distilled rice wine), and soy sauce. *A. oryzae* produces large quantities of a glucoamylase in solid-state cultures (Koji) [168], but the productivity in submerged cultures is much lower [169]. This is because the *A. oryzae* glucoamylase encoding gene, *glaB*, is highly expressed in solid-state cultures but repressed in submerged cultures [170].

In solid-state cultures (Koji-making) during saké brewing, *A. oryzae* strains that produce abundant glucoamylase also produce a lot of tyrosinase, which catalyzes both the *o*-hydroxylation of phenols and the oxidation of diphenols to quinones, followed by spontaneous polymerization of the *o*-quinones to melanins [147]. An unfavorable black saké cake called *Kuro-kasu,* whose color is dependent on melanin. This led to the hypothesis that *A. oryzae glaB* and the tyrosinase-encoding gene (*melO*) are subject to the same regulation in solid-state cultures [149].

To test this hypothesis, expression of the *glaB* and *melO* genes was compared in a β-glucuronidase (GUS) reporter assay. Regulation of *melO*, however, was completely different from that of *glaB*. The *melO* promoter was repressed in solid-state culture, but more highly expressed in submerged culture than were *A. oryzae amyB* [170], *glaA* [171], and modified *agdA* [172, 173], a promoter frequently used in the *A. oryzae* heterologous gene expression system. This finding encouraged us to the *melO* gene promoter to establish a novel expression system in *A. oryzae*. [174].

9.2. Hyper Production System of Glucoamylase Under *MelO* Gene Promoter Control

UV-mediated mutagenesis generated a high glucoamylase-producing mutant (OSI 1117) of *A. oryzae* exhibiting strong melanization in a solid-state cultures. The mutant produced 1.8-fold more glucoamylase and 1.3-fold more tyrosinase than the parent strain and showed higher melanization, indicating that the improvement of glucoamylase production in solid-state cultures through UV-mediated mutagenesis simultaneously leads to higher tyrosinase production and melanization.

Expression of the glucoamylase-encoding gene (*glaB*) and the tyosinase-encoding gene (*melO*) was analyzed using an *E. coli* β-glucuronidase (GUS) reporter-assay. Although no common regulation was found for *melO* and *glaB* expression, the former was greatly enhanced in submerged cultures. Intrestingly, the *melO* gene promoter was about four times stronger for β-glucuronidase production than the powerful promoter *amyB*, *glaA*, and modified *agdA*, previously isolated for industrial heterologous gene expression in *A. oryzae*.

These findings indicated that the *melO* gene promoter would be suitable for hyper-production of heterologous protein in *Aspergillus*. Thus *glaB*-type glucoamylase as the target protein was produced in a submerged cultures of *A. oryzae* under the control of the *melO* gene promoter. Figure 41 shows the purities of both culture broths analyzed by SDS-PAGE after 10 days of cultures [174]. Many impurities are present in the culture broth of dextran-yeast extract-peptone (DPY) medium (Figure 41, Left), whereas only glucoamylase (*glaB*-type) is present in the standard cultural broth (Figure 41, Right). The maximum yield was 0.8 g/l broth, and the total extracellular protein purity was 99%.

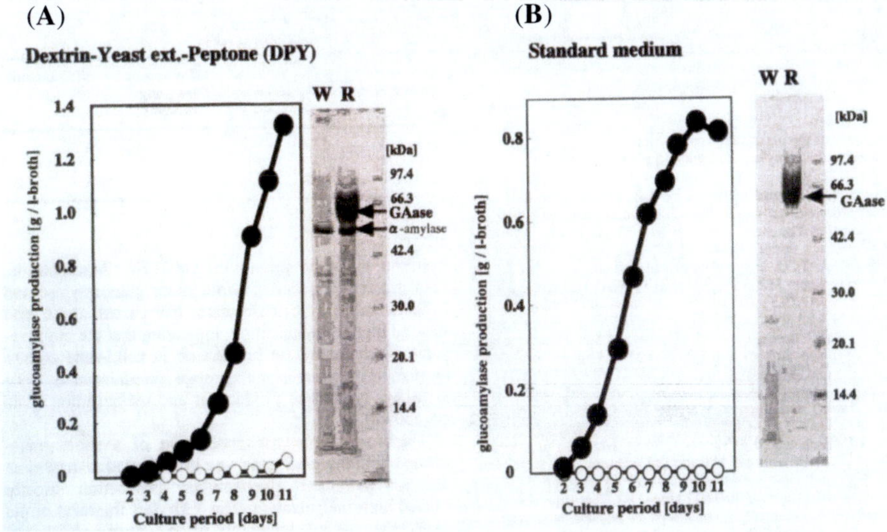

Figure 41. *glaB*-type glucoamylase production under *melO* promoter control in submerged culture of *A. oryzae*. **Left panels**: time course, *black circles* (●) recombinant (NMB1 harbouring one copy of pNMB1), *open circles* (○) wild-type (OSI1002). **Right panels**: SDS-PAGE analysis (10-20% polyacrylamide gel) of 20 μl of each culture broth type after 10 days of batch culture. **A**, Dextrin-yeast extract-peptone (DPY) medium; **B**, standard medium.

(Ishida, H. *et al., Appl. Microbiol. Biotechnol., 57,* 131-137 (2001))

Although, the *melO* expression system was a powerful tool for protein production in submerged cultures, it took a long time to obtain high target protein productivity. Because of marked expression of products after the stationary growth phase, repeated batch cultures were done for the efficient use of mycelia after the stationary growth phase. The experiment shows that repeated batch culture increased glucoamylase production, the maximum being 3.3 g/l broth on the 12th day after culture was begun. Productivity for the 21 days from the beginning of repeated batch cultures was more than 2 g/l broth, and the product purity was 99% as total extracellular protein throughout the culture period.

The importance of this work is in the establishment of both a high-level and high-purity protein overproduction system in *A. oryzae* by use of the *melO* gene promoter [174].

APPENDIX: ENZYMES REFERRED TO IN CHAPTERS 1-9

The enzymes referred to in the chapters of this review are listed in alphabetical order generally under their recommended names as in *ENZYME NOMENCLATURE* (Recommendation (1992) of Nomenclature Committee of the International Union of Biochemistry and Molecular Biology), except in a few cases where trivial names are more frequently used.

EC number

Aorsin 3.4.21.–

Activation of trypsinogen by selectively cleavage of Lys^6- ↓ -Ile bond at pH 4. Specific requirement for dibasic residues at P2 and P1; Boc-Gln-Arg-Arg- ↓ -MCA.

Aspergillopepsin I 3.4.23.18

Hydrolyses proteins with broad specificity. Generally favours hydrophobic residues in P_1 and P_1', but accept Lys in P_1, which leads to activation of trypsinogen. Does not clot milk.

Carboxypeptidase C 3.4.16.5

Other name: *Aspergillus acid carboxypeptidase*

Release of a C-terminal amino acid with broad specificity at pH 3-5. Prefer hydrophobic residues in the P_1 position.

Catechol oxidase 1.10.3.1

Other name: Tyrosinase

Reaction: 2 Catechol + O_2 = 2 1,2-benzoquinone + 2 H_2O

Deuterolysin 3.4.24.39

Other name: *Aspergillus sojae* neutral proteinase II; Microbial neutral proteinase II

Preferential cleavage of bonds with hydrophobic residues in P_1'; also Asn^3- ↓ -Gln and Gly^8- ↓ -Ser bonds in insulin B chain. For calf thymus histone H4, high activity towards peptides next to a pair of basic residues at pH 7.0; Lys-Arg^{17}- ↓ -His, Arg-His^{18}- ↓ -Arg, Arg-Arg^{36}- ↓ -Leu.

Enteropeptidase 3.4.21.9

Other name: Enterokinase

Activation of trypsinogen by selectively cleavage of Lys^6- ↓ -Ile bond.

Glucan 1,4-α-glucosidase 3.2.1.3

Other name: Glucoamylase

Hydrolysis of terminal 1,4-linked α-D-glucose residues successively from non-reducing ends of the chains with release of β-D-glucose.

Mannosyl-oligosaccharide 1,2-α-mannosidase 3.2.1.1.113

Other name: 1,2-α-D-mannosidase

Systematic name: 1,2-α-mannosyl-oligosaccharide-α-D-mannohydrolase

Hydrolysis of the terminal 1,2-linkaged α-D-mannose residues in the oligo-mannose oligosaccharide $Man_9(GlcNAc)_2$.

Monophenol monooxygenase 1.14.18.1

Other name: Tyrosinase

Reaction: L-Tyrosine + L-DOPA + O_2 = L-DOPA + dopaquinone + H_2O

Penicillolysin 3.4.24.–

The rate of clupeine hydrolysis is 60-fold greater than that for casein hydrolysis at pH 7.0. High affinity towards the Pro- ↓ -X, Arg- ↓ -Arg, and hydrophobic residue (P_1)- ↓ -X.

Pepsin A 3.4.23.1

Preferential cleavage: hydrophobic, preferably aromatic residues in P_1 and P_1' positions. Clevages Phe^1-Val, Gln^4-His, $Glu1^3$-Ala, Ala^{14}-Leu, Leu^{15}-Tyr, Tyr^{16}-Leu, Gly^{23}-Phe, Phe^{24}-Phe, and Phe^{25}-Tyr bonds in the B chain of insulin.

Acknowledgments

The author gratefully dedicates this review to the co-workers, for their kind collaborations, as well as their technical assistance, in these studies. The review is compiled in commemoration of our long enzyme research works published in original articles, and the author hopes that our results will contribute to helping others in attaining greater heights.

Profile

Eiji Ichishima was born in 1934. He graduated from the Tokyo University of Agriculture and Technology, Department of Agricultural Chemistry, in 1957. He received his Ph.D. in 1967 from the Tokyo University. In 1957, he started his research career with studies on enzyme chemistry at the Noda Institute for Scientific Research and Central Research Institute of Kikkoman Co., Ltd. He was promoted to associate professor in 1969, and to professor in 1975 at the Tokyo University of Agriculture and Technology. After 1987, he moved to the Tohoku University, Department of Agricultural Chemistry and Applied Biochemistry, as a professor. Between 1997 and 2009, he was a professor at the Soka University, Faculty of Engineering. In 1972, he received the Promotion Award from the Japanese Biochemical Society. He received the Prime Minister's Prize for Invention in 1966, the Technological Achievement Award from the Agricultural Chemical Society of Japan in 1968, the Academic Achievement Award from the Agricultural Chemical Society of Japan in 1997, and the Academic Achievement Award from the Brewing Society of Japan in 2008. He was nominated as Professor Emeritus of Tohoku University and Tokyo University of Agriculture and Technology in 1997. He retired in 2006.

REFERENCES

[1] Ichishima, E. (2007) *Koji (Japanese Malt made by Aspergillus fungus)* (*In Japanese*). Chiyoda-ku, Kudan-kita, Tokyo: Hosei University Press.

[2] Akiyama, H. (2010) *SAKÉ (The Essence of 2000 Years of Japanese Wisdom gained from Brewing Alcoholic Beverages from Rice).* Translated by Inoue, T. Kita-ku, Takinogawa, Tokyo: Brewing Society of Japan INTERNATIONAL EDITION.

[3] Fukushima, D. (1985). Fermented vegetable protein and related foods of Japan and China. *Food Rev. Int.*, *1*, 149-209.

[4] Takamine, J. (1914). Enzymes of *Aspergillus oryzae* and the application of its amyloclastic enzyme to the fermentation industry. *Ind. Eng. Chem.*, *6*, 821-828.

[5] Sakaguchi, K., Iizuka, H. and Yamazaki, S. (1951). A study on blask *Aspergilli*, *Nippon Nogeikagakku Kaishi* (in Japanese, *J. Agric. Chem. Soc. Jpn.*), *24*, 138-142.

[6] Ichishima, E. (1970). Purification and mode of assay for acid proteinase of *Aspergillus saitoi*. In: Perlmann, G. E., and Lorand, L. (Eds.), *Methods in ENZYMOLOGY: PROTEOLYTIC ENZYMES. Vol. XIX*, (pp.397-406). New York: ACADEMIC PRESS.

[7] Ichishima, E. (2004). Aspergillopepsin I. In: Barrett, A.J., Rawlings, N.D., and Woessner, J.F. (Eds.), *HANDBOOK of Proteolytic Enzymes: Vol. 1*, (Second edition, pp.92-99). Theobald's Road, London: ELSEVIER ACADEMIC PRESS.

[8] Machida, M. *et al.* (2005). Genome sequencing and analysis of *Aspergillus oryzae*. *Nature, 438*, 1157-1161.

[9] Tominaga, M., Lee, Y. H., Hayashi, R., Suzuki, Y., Yamada, O., Sakamoto, K., Gotoh, K. and Akita, O. (2006). Molecular analysis of an

inactive aflatoxin biosynthesis gene cluster in *Aspergillus oryzae* RIB strains. *Appl. Environ. Micribol.*, *72*, 484-490.

[10] Tang, J. (2004). Pepsin A. In Barrett, A.J., Rawlings, N.D., and Woessner, J.F. (Eds.), *HANDBOOK of Proteolytic Enzymes: Vol. 1*, (Second edition, pp.19-28). Theobald's Road, London: ELSEVIER ACADEMIC PRESS.

[11] Kay, J. (1985). Aspartic proteinases and their inhibitors. *Biochem. Soc. Trans.*, *13*, 1027-1029.

[12] Schechter, I. and Berger, A. (1967). On the size of the active site in proteases. I. Papain. *Biochem. Biophys. Res. Commun.*, *27*, 157-162.

[13] Tang, J., James, M. N. G., Hsu, I. N., Jenkins, J. A. and Blundell, T. L. (1978). Structural evidence for gene duplication in the evolution of the acid proteases. *Nature*, *271*, 618-621.

[14] James, M. N. G. and Sielecki, A. (1983). Structure and refinement of penicillopepsin at 1.8 Å resolution. *J. Mol. Biol.*, *163*, 299-361.

[15] Suzuki, F., Murakami, K. and Nakamura, Y. (2004). Renin. In Barrett, AJ, Rawlings, ND, and Woessner, JF. (Eds.), *HANDBOOK of Proteolytic Enzymes: Vol. 1*, (Second edition, pp. 54-61). Theobald's Road, London: ELSEVIER ACADEMIC PRESS.

[16] Foltmann, B. (2004). Chymosin. In: Barret, AJ, Rawlings, ND, and Woessner, JF. (Eds.), *HANDBOOK of Proteolytic Enzymes: Vol. 1*, (Second edition, pp. 29-32). Theobald's Road, London: ELSEVIER ACADEMIC PRESS.

[17] Ballou, C. E. (1990). Isolation, characterization, and properties of *Saccharomyces cerevisiae mann* mutants with nonconditional protein glycosylation defects. In Goeddel, DV, editor. *Methods in ENZYMOLOGY, Vol. 185*, ACADEMIC PRESS, San Diego, Calif., pp.440-470.

[18] Tanaka, N., Takeuchi, M.. and Ichishima, E. (1977). Purification of an acid proteinase from *Aspergillus saitoi* and characterization of peptide bond specificity. *Biochim. Biophys. Acta*, *485*, 406-416.

[19] Majima, E., Oda, K., Murao, S. and Ichishima, E. (1988). Comparative study on the specificity of several fungal aspartic and acidic proteinases towards thetetradecapeptide of a rennin substrate. *Agric. Biol. Chem.*, *52*, 787-793.

[20] Gabeloteau, C. and Desnuelle, P. (1960). On the activation of beef trypsinogen by a crystallized protease of *Aspergillus saitoi. Biochim. Biophys. Acta*, *42*, 230-237

[21] Abita, J. P., Delaage, M., Lazdunski, M. and Savrda, J. (1969). The mechanism of activation of trypsinogen. The role of the four *N*-terminal aspartyl residues. *Eurp. J. Biochem.*, *8*, 314-324.

[22] Hofmann, T. (2004). Penicillopepsin. In Barrett, A.J., Rawlings, N.D., and Woessner, J.F. (Eds.), *HANDBOOK of Proteolytic Enzymes: Vol. 1*, (Second edition, pp.99-104). Theobald's Road, London: ELSEVIER ACADEMIC PRESS.

[23] Sadler, J. E. (2004). Enteropeptidase. In: Barrett, A.J., Rawlings, N.D., and Woessner, J.F. (Eds.), *HANDBOOK of Proteolytic Enzymes: Vol. 2*. (Second Ed. pp.1513-1517). Theobald's Road, London: ELSEVIER ACADEMIC PRESS.

[24] Morihara, K. and Oka, T. (1973). Comparative specificity of microbial acid proteinases for synthetic peptides. 3. Relationship with their trypsinogen activating ability. *Arch. Biochem. Biophys.*, *157*, 561-572.

[25] James, M. N. G., Sielecki, A. R. , Salituro, F., Rich, D. H. and Hofmann, T. (1982). Conformational flexibility in the active site site of aspartyl proteinases revealed by a pepstatin fragment binding to penicillopepsin. *Proc. Nat. Acad. Aci. USA*, **79**, 6137-6141.

[26] James, M. N. G., Sielecki, A.R. and Hofmann, T. (1985). X-ray diffraction studies of penicillopepsin and its complex: the hydrophobic Mechanism. In Kostka, V. (Ed.), *Aspartic Proteinases and Their Inhibitors*. (pp.163-177). Berlin, Walter de Gruyter.

[27] Blundell, T. L., Cooper, J., Foundling, S. I., Jones, D. M., Atrash, B. and Szelk, M. (1987). On the rational design of renin inhibitors: X-ray studies of aspartic proteinase complexed with trasition-state analogues. *Biochemistry*, *26*, 5585-5590.

[28] Suguna, K., Bott, R. R., Padlan, E. A., Subramanian, E., Sheriff, S., Cohen, G. H. and Davies, D. R. (1987). Structure and refinement at 1.8 Å resolution of the aspartic proteinase from *Rhizopus chinensis*. *J. Mol. Biol.*, *196*, 877-900.

[29] Shintani, T. and Ichishima, E. (1994). Primary structure of aspergillopepsin I deduced from nucleotide sequence of the gene and aspartic acid-76 is an essential active site of the enzyme for trypsinogen activation. *Biochim. Biophys. Acta*, *1204*, 257-264.

[30] Shintani, T., Kobayashi, M. and Ichishima, E. (1996). Characterization of the S_1 subsite specificity of aspergillopepsin I by site-directed mutagenesis. *J. Biochem. (Tokyo)*, *120*, 974-981.

[31] Lowther, W. T., Majer, P. and Dunn, B. M. (1995). Engineering the substrate specificity of rhizopuspepsin: the role of Asp 77 of fungal

aspartic proteinase in facilitating the cleavage of oligopeptide substrates with lysine in P_1. *Protein Sci.*, *4*, 689-702.

[32] Beppu, T., Park, Y.-N., Aikawa, J., Nishiyama, M. and Horinouchi, S. (1995). Tyrosine 75 on the flap contributes to enhance catalytic efficiency of a fungal aspartic proteinase, *Mucor pusillus* pepsin. In: *Aspartic Proteinase: Structure, Function, Biology, and Biomedical Implications*. Takahashi, T. (Ed.), (pp. 559-563). New York: Plenum Press.

[33] Dhanaraj, V., Dealwis, C. W., Frazao, C., Badasso, M., Sibanda, B. L, Tickle, I. J., Cooper, J. B., Driessen, H. P., Newman, M., Aguilar, C., Wood, S. P., Blundell, T. L., Hobart, P. M., Geophegan, K. F., Ammirati, M. J., Danley, D. E., O'Connor, B. A. and Hoover, D. J. (1992). X-ray analyses of peptide-inhibitor complexes define the structural basis of specificity for human and mouse renins. *Nature*, *357*, 466-472.

[34] Fujinaga, M., Chernaia, M. M., Tarasova N. I., Mosimann, S. C. and James, M. N. G. (1995). Crystal structure of human pepsin and its complex with pepstatin. *Protein Sci.*, *4*, 960-972

[35] Shintani, T., Nomura, K. and Ichishima, E. (1997). Engineering of porcine pepsin: Alteration of S_1 substrate specificity of pepsin to those of fungal aspartic proteinase by site-directed mutagenesis. *J. Biol. Chem.*, **272**, 18855-18861.

[36] Lin, X., Wong, R. N. S. and Tang, J. (1989). Synthesis, purification, and active site mutagenesis of recombinant porcine pepsinogen, *J. Biol. Chem.*, *264*, 4482-4489.

[37] Tang, J., Sepulveda, P., Marcinszyn, J., Jr., Chen, K. C. S., Huang, W.-Y., Tao, N., Liu, D. and Lanier, J. P. (1973). Amino acid sequence of porcine pepsin. *Proc. Natl. Acad. Sci. USA*, *70*, 3437-3439.

[38] Tsukagoshi, N., Ando, Y., Tomita, Y., Uchida, R., Takemura, T., Sasaki, T., Yamagata, H., Udaka, S., Ichihara, Y. and Takahashi, K. (1988). Nucleotide sequence and expression in *Escherichia coli* of cDNA of swine pepsinogen: involvement of the amino-terminal porsine of the activation peptide segment in restoration of the functional protein. *Gene*, *65*, 285-292.

[39] Green, B. N., Jones, A. T. and Roberts, N. B. (1996). Electrospray mass spectrometric evidence for the occurrence of two major variants in native pig pepsin A. *Biochem.J.*, *313*, 241-244.

[40] Kageyama, T. and Takahashi, K. (1983). Occurrence of two different pathways in the activation of porcine pepsinogen to pepsin. *J. Biochem. (Tokyo)*, *93*, 743-754.

[41] Pohl, J. and Dunn, B. M. (1988). Secondary enzyme-substrate interactions: kinetic evidence for ionic interactions between substrate side chains and the pepsin active site. *Biochemistry*, *27*, 4827-4834.

[42] Takahashi, K. (1995). Proteinase A from *Aspergillus niger*. In Barrett, A. J. (Ed). *Methods in ENZYMOLOGT, Vol. 248. Proteolytic Enzymes: Aspartic and Metallo Peptidases*, (pp. 146-155.). New York; ACADEMIC PRESS.

[43] Murao, S., Oda, K. and Matsushita, Y. (1972). New acid protease from *Scytalidium lignicolum., Agric. Biol. Chem.*, *36*, 1647-1650.

[44] Oda, K., Takahashi, T., Tokuda, Y., Shibano, Y. and Takahashi, S. (1994). Cloning, nucleotide sequence, and expression of an isovalery pepstatin-insensitive carboxyl proteinase gene from *Pseudomonas* sp. 101., *J. Biol. Chem.*, *269*, 26518-26524.

[45] Oda, K., Nakajima, T., Terashita, T., Suzuki, K. and Murao, S. (1987). Purification and properties of an S-PI (pepstatin Ac)-insensitive carboxyl proteinase from *Xanthomonas* sp. bacterium. *Agric. Biol. Chem.*, *51*, 3073-3080.

[46] Oda, K., Ito, M., Uchida, K., Shibano, Y., Fukuhara, K. and Takahashi, S. (1996). Cloning and expression of an isovaleryl pepstatin-insensitive carboxyl proteinase gene from *Xanthomonas* sp. T-22., *J. Biochem. (Tokyo)*, *120*, 564-572.

[47] Sleat, D. E., Donnelly, R. J., Lackland, H., Liu, C. –G., Sohar, I., Pullarkat, R. K. and Lobel, P. (1997). Association of mutations in a lysosomal protein with classical lateinfantile ceroid lipofuscinosis. *Science*, *277*, 1802-1805.

[48] Wisniewski, K. E., Kaczmarski, A., Kida, E., Connell, F., Kaczmaeski, W., Michaelewski, M. P., Moroziewicz, D. N. and Zhong, N. (1999). Reevaluation of neuronal ceroid lipofuscinoses: atypical juvenile onset may be the result of CLN2 mutations., *Mol. Genet. Metab.*, *66*, 248-252.

[49] Junaid, M. A., Wu, G. and Pullarkat, R. K. (2000). Purification and characterization of bovine brain lysosomal pepstatin-insensitive proteinase, the gene product deficient in the human late-infantile neuronal ceroid lipofuscinosis. *J. Neurochem.*, *74*, 287-294.

[50] Rawlings, N. D. and Barrett, A. J. (1999). Tripeptidyl-peptidase I is apparently the CLN2 protein absent in classical late-infantile neuronal ceroid lipofuscinosis. *Biochim. Biophys. Acta*, *1429*, 496-500.

[51] Ezaki, J., Takeda-Ezaki, M., Oda, K. and Kominami, E. (2000). Characterization of endopeptidase activity of tripeptidyl peptidase-I/CLN2 protein which is deficient in classical late infatile neuronal ceroid lipofuscinosis. *Biochem. Biophys. Res.Commun.*, *268*, 904-908.

[52] Lin, L., Sohar, I., Lackland, H. and Lobel, P. (2001). The human CLN2 protein/tripeptidyl-peptidase I is a serine protease that autoactivates at acidic pH., *J. Biol. Chem.*, *276*, 2249-2255.

[53] Wlodawer, A., Li, M., Dauter, Z., Gustchina, A., Uchida, K., Oyama, H., Dunn, B. M. and Oda, K. (2001). Carboxyl proteinase from *Pseudomonas* defines a novel family of subtilisin-like enzymes. *Natl. Struct. Biol.*, *8*, 442-446.

[54] Rawlings, N. D., O'Brien, E. A. and Barrett, A. J. (2002). MEROPS: the protease database. *Nucleic Acids Res.*, *30*, 343-346.

[55] Lee, B. R., Furukawa, M., Yamashita, K., Kanasugi, Y., Kawabata, C., Hirano, K., Ando, K. and Ichishima, E. (2003). Aorsin, a novel serine proteinase with trypsin-like specificity at acidic pH. *Biochem. J.*, *371*, 541-548.

[56] Oda, K., Fukuda, Y., Murao, S., Uchida, K. and Kainosho, M. (1989). A novel proteinase inhibitor, tyrostatin, inhibiting some pepstatin-insensitive carboxyl proteinases. *Agric. Biol. Chem.*, *53*, 405-415.

[57] Oyama, H., Abe, S., Ushiyama, S., Takahashi, S. and Oda, K. (1999). Identification of catalytic residues of pepstatin-insensitive carboxyl proteinases from prokaryotes by site-directed mutagenesis. *J. Biol. Chem.*, *274*, 27815-27822.

[58] Sekine, H. (1973). Neutral proteinases II of *Aspergilus sojae*: an enzyme specifically active on protamine and histone. *Agric. Biol. Chem.*, *37*, 1765-1767.

[59] Tatsumi, H. (2004). Deuterolysin. In Barrett, A. J., Rawlings, N. D., and Woessner, J. F. (Eds.), *HANDBOOK of Proteolytic Enzymes: Vol. 1*, (Second edition, pp. 786-788). Theobald's Road, London: ELSEVIER ACADEMIC PRESS.

[60] Tatsumi, H., Murakami, S., Tsuji, R. E., Ishida, Y., Murakami, K., Masaki, A.,Kawabe, H., Arimura, H., Nakano, E. and Motai, H. (1991). Cloning and expression in yeast of a cDNA clone encoding *Aspergillus oryzae* neutral protease II, a unique metalloprotease. *Mol. Gen. Genet.*, *228*, 97-103.

[61] Tatsumi, H., Ikegaya, K., Murakami, S., Kawabe, H., Nakano, E. and Motai, H. (1994). Elucidation of the thermal stability of the neutral

proteinase II from *Aspergillus oryzae. Biochim. Biophys. Acta*, *1208*, 179-185.

[62] Hooper, N.M. (1994). Families of zinc metalloproteases. *FEBS Lett.*, *354*, 1-6.

[63] Rawlings, N.D. and Barrett, A. J. (1995). *Evorutionary families of metallopeptidases*. In: Barrett, A. J. (Ed.), *Methods in Enzymology, Vol. 248,* (pp. 183-228). San Diego: ACADEMIC PRESS.

[64] Fushimi, N., Ee, C. E., Nakajima, T. and Ichishima, E. (1999). Aspzincin, a family of metalloendopeptidase with a new zinc-binding motif. *J. Biol. Chem.*, *274*, 24195-24201.

[65] McAuley, K. E., Jia-Xing, Y., Dodson, E. J., Lehmbeck, J., Ostergaard, P. R. and Wilson, K. S. (2001). A quick solution: *ab initio* structure determination of a 19 kDa metalloproteinase using ACORN. *Acta Crystallogr. D.*, *57*, 1571-1578.

[66] Hori, To, Kumasaka, T., Yamamoto, M., Nonaka, N., Tanaka, N., Hashimoto, Y., Ueki, T. and Takoi, K. (2001). Structure of a new 'aspzincin' metalloendopeptidase from *Grifola frondosa*: implications for the catalytic mechanism and substrate specificity based on several different crystal forms. *Acta Crystallogr. D.*, *57*, 361-368.

[67] Doi, Y., Lee, B. R., Ikeguchi, M., Ohoba, Y., Ikoma, T., Terao-Kubota, S., Yamauchi, S., Takahashi, K. and Ichishima, E. (2003). Substrate specificities of deuterolysin from *Aspergillus oryzae* and electron paramagnetic resonance measurement of cobalt-substituted deuterolysin. *Biosci. Biotechnol. Biochem.*, *67*, 264-270.

[68] Ichishima, E., Yamaguchi, M., Yano, H., Yamagata, Y. and Hirano, K. (1991). Specificity of a new metalloproteinase from *Penicillium citrinum. Agric. Biol. Chem.*, *55*, 2191-2193.

[69] Yamaguchi, M., Hanzawa, S., Hirano, K., Yamagata, Y. and Ichishima, E. (1993). Specificity and molecular properties of penicillolysin, a new metalloproteinase from *Penicirium citrinum. Phytochemistry*, *33*, 1371-1321.

[70] Matsumoto, K., Yamaguchi, M. and Ichishima, E. (1994). Molecular cloning and nucleotide sequence of the complementary DNA for penicillolysin gene, *plnC*, an 18 kDa metalloendopeptidase gene from *Penicillium citrinum. Biochim. Biophys. Acta*, *1218*, 469-472.

[71] Ichishima, E. (2004). Penicillolysin. In Barrett, A.J., Rawlings, N.D., and Woessner, J.F. (Eds.), *HANDBOOK of Proteolytic Enzymes: Vol. 1*, (Second edition, pp.784-786). Theobald's Road, London: ELSEVIER ACADEMIC PRESS.

[72] Doi, Y., Akiyama, H., Yamada, Y., Ee, C. E., Lee, B. R., Ikeguchi, M. and Ichishima, E. (2004). Thermal stabilization of penicillolysin, a thermolabile 19 kDa Zn^{2+}-protease, obtained by site-directed mutagenesis. *Protein Eng. Des. Select.*, *17*, 261-266.

[73] Price, N. C. and Stevens, L. (1999). *Fundamentals of Enzymology: Enzyme turnover* (Third edition, pp.370-399). Great Clarendon, Oxford: Oxford University Press.

[74] Matthews, B. W., Nicolson, H. and Becktel, W. J. (1987). Enhanced protein thermostability from site-directed mutagenesis that decrease the entropy of unfolding. *Proc. Natl. Acad. Sci. USA*, *84*, 6663-6666.

[75] Suzuki, Y. (1989). A general principle of increasing protein thermostability. *Proc. Jpn. Acad.*, **65**, 146-148.

[76] Suzuki, Y., Oishi, K., Nakano, H. and Nagayama, T. (1987). A strong correlation between the increase in number of proline residues and the rise in thermostability of five *Bacillus* oligo 1,6-glucosidases. *Appl. Microbiol. Biotechnol.*, *26*, 546-551.

[77] Auld, D. S. (2004). Carboxypeptidase A. In Barrett, A.J., Rawlings, N. D., and Woessner, J. F. (Eds.), *HANDBOOK of Proteolytic Enzymes: Vol. 1*, (Second edition, pp. 813-821). Theobald's Road, London: ELSEVIER ACADEMIC PRESS.

[78] Aviles, F. X. and Vendrell, J. (2004). Carboxypeptidase B. In Barrett, A.J., Rawlings, N.D., and Woessner, J. F. (Eds.), *HANDBOOK of Proteolytic Enzymes: Vol. 1*, (Second edition, pp. 831-833). London, ELSEVIER ACADEMIC PRESS.

[79] Ichishima, E. (1972). Purification and characterization of a new type of acid aroxypeptidase from *Aspergillus. Biochim. Biophys. Acta*, *258*, 274-288.

[80] Ichishima, E. and Arai, T. (1973). Specificity and mode of action of acid carboxypeptidase from *Aspergillus saitoi. Biochim. Biophys. Acta*, *293*, 444-450.

[81] Chiba, Y., Yamagata, Y., Nakajima, T. and Ichishima, E. (1992). A new high-mannose type-*N*-linked oligosaccharide from *Aspergillus* carboxypeptidase. *Biosci. Biotechnol. Biochem.*, *56*, 1371-1372.

[82] Chiba, Y., Yamagata, Y., Iijima, S., Nakajima, T. and Ichishima, E. (1993). The carbohydrate moiety of the acid carboxypeptidase from *Aspergillus saitoi*. *Curr. Microbil.*, *27*, 281-288.

[83] Takaki, A., Yoshitake, S., Ishiguro, M. and Funatsu, M. (1972). Anticoagulant peptide obtained from fibrinogen degradation products by

plasmin. II. Sequence determination of the peptide. *Proc. Japan Acad.*, *48*, 525-538.

[84] Chiba, Y., Nieda, Y., Nakajima, T. and Ichishima, E. (1991) Unique enzymatic properties of alpha-amylase-III from suspension-cell cultured rice cells. *Agric. Biol. Chem.*, *55*, 901-902.

[85] Yanagida, N., Uozumi, T. and Beppu, T. (1986). Specific excretion of *Serratia marcescens* protease through the outer membrane of *Escherichia coli*. *J. Bacteriol.*, *166*, 937-944.

[86] Takeuchi, M. and Ichishima, E. (1989). Action of serine carboxypeptidases from *Aspergillus saitoi* on carboxyterminal amidated peptides. *Agric. Biol. Chem.*, *53*, 2301-2306.

[87] Takeuchi, M., Takagi, Y., Ebisui, R., Toyama, T. and Ichishima, E. (1989). Mode of action of fluorogenic substrates of acid carboxypeptidases from *Aspergillus*. *Agric. Biol. Chem.*, *53*, 1177-1178.

[88] Chiba, Y., Midorikawa, T. and Ichishima, E. (1995). Cloning and expression of the carboxypeptidase gene from *Aspergillus saitoi* and determination of the catalytic residues by site-directed mutagenesis. *Biochem. J.*, *308*, 405-409.

[89] Mortensen, U. H., Olesen, K. and, Breddam, K. (2004). Serine carboxypeptidsase C including carboxypeptidase Y. In Barrett, A.J., Rawlings, N.D., Woessner, J.F. (Eds.), *HANDBOOK of Proteolytic Enzymes: Vol. 2*, (Second edition, pp. 1919-1923). Theobald's Road, London: ELSEVIER ACADEMIC PRESS.

[90] Arai, S., Yamashita, M., Kato, H. and Fujimaki, M. (1970). Applying proteolytic enzymes on soybean. Part V. A nondialyzable bitter peptide in peptic hydrolyzate of soybean protein and its bitterness in relation to the chemical structure. *Agric. Biol. Chem.*, *34*, 729-738.

[91] Umetsu, H., Matsuoka, H., and Ichishima, E. (1983). Debittering mechanism of bitter peptides from milk casein by wheat arboxypeptidase. *J. Agric. Food Chem.*, *31*, 50-53.

[92] Hernandez, L. M., and Ballou, L., Alvarado, E., Gillece-Castro, B. L., Burlingame, A. L. and Ballou, C. E. (1989). A new *Saccharomyces cerevisiae mnn* mutant N-linked ologisaccharide structure. *J. Biol. Chem.*, *264*, 11849-11856.

[93] Cohen, R. E., Zhnag, W. and Ballou, C. E. (1982). Effects of mannoproteins in *Saccharomyces cerevisiae* core oligosaccharide structure. *J. Biol. Chem.*, *257*, 5730-5737.

[94] Ichishima, E., Arai, M., Shigematsu, Y., Kumagai, H. and Sumida-Tanaka, R. (1981). Purification of an acidic α-D-mannosidase from

Aspergillus saitoi and specific cleavage of 1,2-α-D-mannosidic linkage in yeast mannan. *Biochim. Biophys. Acta,658*, 45-53.

[95] Yamashita, K., Ichishima, E., Arai, M. and Kobata, A. (1980). An α-mannosidase purified from *Aspergillus saitoi* is specific for α1,2 linkages. *Biochem. Biophys. Res.Commun.*, *96*, 1335-1342.

[96] Trimble, R. B. and Atkinson, P. H. (1986). Structure of yeast external invertase $Man_{8-14}GlcNAc$ processing intermediates by 500-megahertz ^{1}H NMR spectroscopy. *J. Biol. Chem.*, *261*, 9815-9824.

[97] Ito, S., Yamashita, K., Spiro, R. G. and Kobata, A. (1977). Structure of carbohydrate moiety of a unit A glycopeptide of calf thyroglobulin. *J. Biochem.*, *81*, 1621-1631 (1977).

[98] Öckerman, P. A. (1967). A generalized storage disdorder resembling Hurler's syndrome. *Lancet II*, 239-241.

[99] Öckerman, P. A., Autio, S. and Norder, N. E. (1973) Diagnosis of mannosidosis. *Lancet 1*, 207-208.

[100] Champion, M. J. and Shows, T. B. (1977) Mannosidosis: Assignment of the lysosomal α-mannosidase B gene to chromosome 19 in man. *Proc. Natl. Acad. Sci. USA, 74*, 2968-2972.

[101] Moremen, K.W., Trimble, R. B. and Herscovics, A. (1994). Glycosidases of the asparagines-linked oligosaccharide processing pathway. *Glycobiology*, *4*, 113-125.

[102] Schneikert, J. and Herscovics, A. (1995). Two naturally occurring mouse α-mannosidase IB cDNA clones differ in three point mutations. Mutation of Phe592 to Ser592 is sufficient to abolish enzyme activity. *J. Biol. Chem.*, *270*, 17736-17740.

[103] NC-IUBMN, In: Webb, E. C. Editor. *Enzyme Nomenclature 1992*. ed. San Diego, California: ACADEMIC PRESS, INC. (1992).

[104] Yoshida, T., Inoue, T. and Ichishima, E. (1993). 1,2-α-D-mannosidase from *Penicillium citrinum*: molecular and enzymatic properties of two isoenzymes. *Biochem. J.*, *290*, 349-354.

[105] Yoshida, T., Maeda, K., Kobayashi, M. and Ichishima, E. (1994). Chemical modification of *Penicillium* 1,2-α-D-mannosidase by water-soluble carbodi-imide: identification of a catalytic important aspartic acid residue. *Biochem. J.*, *303*, 97-103.

[106] Ichishima, E., Ito, Y. and Takeuchi, M. (1985). 1,2-α-D-mannosidase from a wood-rotting Basidiomycete, *Pycnoporus sanguineus*. *Phytochemistry*, *24*, 2835-2837.

[107] Inoue, T., Yoshida, T, and Ichishima, E. (1995). Molecular cloning and nucleotide sequence of the 1,2-α-D-mannosidase gene, *msdS*, from

Aspergillus saitoi and expression of the gene in yeast cells. *Biochim. Biophys. Acta*, *1253*, 141-145.

[108] Yoshida, T. and Ichishima, E. (1995). Molecular cloning and nucleotide sequence of the genomic DNA for 1,2-α-D-mannosidase gene, *msdC* from *Penicillium citrinum. Biochim. Biophys. Acta*, *1263*, 159-162.

[109] Jelinek-Kelly, S. and Herscovics, A. (1988). Glycoprotein biosynthesis in *Saccharomyces cervisiae*. Purification of the α-mannosidase which removes one specific mannose residue from $Man_9GlcNAc$. *J. Biol. Chem.*, *263*, 14757-14763.

[110] Lipari, F. and Herscovics, A. (1966). Role of the cysteine residues in the 1,2-α-D-mannosidase involved in N-glycan biosynthesis in *Saccharomyces cerevisiae. J. Biol. Chem.*, *271*, 27615-27622.

[111] Herscovics, A., Schneikert, J., Athanassiadis, A. and Moremen, K. W. (1994). Isolation of a mouse golgi mannosidase cDNA, a member of a gene fairy conserved from yeast to mammals. *J. Biol. Chem.*, *269*, 9864-9871.

[112] Ichishima, E., Taya, N., Ikeguchi, M., Chiba, Y., Nakamura, M., Kawabata, C., Inoue, T., Takahashi, K., Minetoki, T., Ozeki, K., Kumagai, C., Gomi, K., Yoshida, T. and Nakajima, T. (1999). Molecular and enzymatic properties of recombinant 1,2-α-mannosidase from *Aspergillus saitoi* overexpressed in *Aspergillus oryzae* cells. *Biochem. J.*, *339*, 589-597.

[113] Fujita, A., Yoshida, T. and Ichishima, E. (1997). Five crucial carboxyl residues of 1,2-α-D-mannosidase from *Aspergillus saitoi* (*A. phoenicis*), a food microorganism, are identical by site-directed mutagenesis. *Biochem. Biophys. Res. Commun.*, *238*, 779-783.

[114] Tatara, Y., Lee, B. R., Yoshida, T., Takahashi, K. and Ichishima, E. (2003). Identification of catalytic residues of Ca^{2+}-independent 1,2-α-D-mannosidase from *A. saitoi* by site-directed mutagenesis. *J. Biol. Chem.*, *278*, 25289-25294.

[115] Lobsanov, Y. D., Vallee, F., Imberty, A., Yoshida, T., Yip, P., Herscovics, A. and Howell, P. L. (2001). Structure of *Penicillium citrinum* α1,2-mannosidase reveals the basic differences in specificity of the endoplasmic reticulum and Golgi class I enzymes. *J. Biol. Chem.*, *277*, 5620-5630.

[116] Miyake, H., Otsuka, C., Nishimura, S. and Nitta, Y. 1 (2002). Catalytic mechanism of β-amylase from *Bacillus cereus* var. *mycoides*: Chemical rescue of hydrolytic activity for a catalytic site mutant (Gku367→Ala) by azide. *J. Biochem.* (*Tokyo*), *131*, 587-591.

[117] Koivula, A., Ruohonen, L., Wohlfahrt, G., Reinkainen, T., Teeri, T. T., Piens, K., Claeyssens, M., Weber, M., Vasella, A., Becker, D., Sinnott, M. L., Zou, J. Y., Kleywegt, G. J., Szardenings, M., Ståhberg, J. and Jones, T. A. (2002). The active site of cellobiohydrolase Ce16A from *Trichoderma reesei*: The roles of aspartic acids D221 and D175. *J. Am. Chem. Soc.*, *124*, 10015-10024.

[118] Koivula, A., Reinikainen, T., Ruohonen, L., Valkeajarvi, A., Claeyssens, M., Teleman, O., Kleywegt, G. T., Szardenings, M., Rouvinen, J., Jones, T. A. and Teeri, T. T. (1996). The active site of *Trichoderma reesei* cellobiohydrolase II: the role of tyrosine 169. *Protein Eng.*, *9*, 691-699.

[119] Zou, J., Kleywegt, G. J., Stahlberg, J., Dringuez, H., Nerinckx, W., Claeyssens, M., Koivula, A., Teeri, T. T. and Jones, T. A. (1999). Crystallographic evidence for substrate ring distortion and protein conformational changes during catalysis in cellobiohydrolase Cel6A from *Trichoderma reesei*. *Structure* (*Lond.*), *7*, 1035-1045.

[120] Hasegawa, K., Kubota, M. and Matsuura, Y. (1999). Role of catalytic residues in α-amylases as evidenced by the structures of the product-complexed mutants of a maltotetraose-forming amylase. *Protein Eng.*, *12*, 819-824.

[121] Gonzalez, D. S., Karaveg, K., Vandersall-Nairn, A.S., Lal., A. and Moremen. K. W. (1999). Identification, expressetion, and characterization of a cDNA encoding human endoplasmic reticulum mannosidase I, the enzyme that catalyzes the first mannose trimming step in mammalian Asn-linked oligosaccharide biosynthesis. *J. Biol. Chem.*, *274*, 21375-21386.

[122] Schutzbach, J. S. and Forsee, W. T. (1990). Calcium ion activation of rabbit liver α1,2-mannosidase. *J. Biol. Chem.*, *265*, 2546-2549.

[123] Oku, H., Hase, S., and Ikenaka, T. (1991). Purification and characterization of neutral α-mannosidase that is activated by Co^{2+} from Japanese quail oviduct. *J. Biochem.* (*Tokyo*), *110*, 29-34.

[124] Yamashiro, K., Itoh, H., Yamagishi, M., Natsuka, S., Mega, T. and Hase, S. (1997). Purification and characterization of neutral α-mannosidase from hen oviduct: Studies on the activation mechanism of Co^{2+}. *J. Biochem.* (*Tokyo*), *122*, 1174-1181.

[125] Lipari, F. and Herscovics, A. (1994). Production, purification and characterization of recombinant yeast processing α1,2-mannosidase. *Glycobiology*, *4*, 697-702.

[126] Jelinek,-Kelly, S. and Herscovics, A. (1988). Glycoprotein biosynthesis in *Saccharomyces cerevisiae*. *J. Biol. Chem.*, *263*, 14757-14763.

[127] Lubas, W. A. and Spiro, R. G. (1988). Evaluation of the role of rat liver golgi endo-α-D-mannosidase in processing N-linked oligosaccharides. *J. Biol. Chem.*, *263*, 3990-3998.

[128] Lipari, F. and Herscovics, A. (1999). Calcium binding to the class I alpha-1,2-mannosidase from *Saccharomyces cerevisiae* occurs outside the EF hand motif. *Biochemistry*, *38*, 1111-1118.

[129] Tatara, Y., Yoshida, T. and Ichishima, E. (2005). A single free cysteine residue and disulfide bond contribute to the thermostability of *Aspergillus saitoi* 1,2-α-D-mannosidase. *Biosci. Biotechnol. Biochem.*, *69*, 2101-2108.

[130] Dobson, C. M. (2003). Protein folding and misfolding. *Nature*, *426*, 884-890.

[131] Sitia, R. and Braakman, I. (2003). Quality control in the endoplasmic reticulum protein factory. *Nature*, *426*, 891-894.

[132] Goldberg, A. L. (2003). Protein degradation and protection against misfolded or damaged proteins. *Nature*, *426*, 859-899.

[133] Lehman, M. A. (2001). Oligosaccharide-based information in endoplasmic reticulumn quality control and other biological system. *J. Biol. Chem.*, *276*, 8623-8626.

[134] Kyte, J. and Doolittle, R. F. (1982). A simple method for displaying the hydropathic character of a protein. *J. Mol. Biol.*, *157*, 105-132.

[135] Chiba, Y., Suzuki, M., Yoshida, S., Yoshida, A., Ikenaga, H., Takeuchi, M., Jigami, Y. and Ichishima, E. (1998). Production of human compatible high mannose-type ($Man_5GlcNAc_2$) sugar chains in *Saccharomyces cerevisiae*. *J. Biol. Chem.*, *273*, 26298-26304.

[136] Garcia-Borron, J. C. and Solano, F. (2002). Molecular anatomy of tyrosinase and its related proteins: beyond the histidine-bound metal catalytic center. *Pigment Cell. Res.*, *15*, 162-173

[137] Kobayashi, T., Vieira, W. D., Potterf, B., Sakai, C., Imokawa, G. and Hearing, V. J. (1995) Modulation of melanogenic protein expression during the switch from eu- to pheomelanogenesis. *J. Cell. Sci.*, *108*, 2301-2309.

[138] Olivers, C., Jimenez-Cervantes, C., Lozano, J. A., Solano, F. and Garcia-Borron, J. C. (2001) The 5,6-dihydroxyindole-2-carboxylic acid (DHICA) oxidase activity of human tyrosinase. *Biochem. J.*, *354*, 131-139.

[139] Riley, P. A. (1997) Melanin. *Int. J. Biochem. Cell Biol.*, *29*, 1235-1239.

[140] Gomez, B. L. and Nosanchuk, J. D. (2003) Melanin and fungi. *Curr. Opin. Infect. Dis.*, *16*, 91-96.

[141] Henson, J. M., Butler, M. J. and Day, A. W. (1999) The dark side of the mycelium: melanins of phytopathogenic fungi. *Annu. Rev. Phytopathol.*, *37*, 447-471

[142] Adman, E. T. (1991). Copper protein structures. *Adv. Protein Chem.*, *42*, 145-197.

[143] Haneda, K., Watanabe, S. and Tanaka, I. (1973) Production of L-3,4-dihydroxyphenylalanine from L-tyrosine by microorganisms. *J. Ferment. Technol.*, *51*, 398-407.

[144] Haneda, K. and Tanaka, I. (1974) Activation of the enzyme of *Aspergillus oryzae* catalyzing the conversion of L-tyrosine to 3,4-dihydroxy-L-phenylalanine. *J. Ferment. Technol.*, *52*, 106-113.

[145] Sih, C. J., Foss, P., Rosazza, J. and Lemberger, M. (1969) Microbiological synthesis of L-3,4-dihydrooxyphenylalanine. *J. Amer. Chem. Soc.*, *91*, 6204.

[146] Haneda, K. and Tanaka, I. (1974) Activation of the enzyme of *Aspergillus oryzae* catalyzing the conversion of L-tyrosine to 3,4-dihydroxy-L-phenylalanine. *J. Ferment. Technol.*, *52*, 106-113.

[147] Ichishima, E., Maeba, H., Amikura, T. and Sakata, H. (1984). Multiple forms of protyrosinase from *Aspergilus oryzae* and their mode of activation at pH 3.0. *Biochim. Biophys. Acta*, *786*, 25-31.

[148] Tanaka, T., Takeuchi, M. and Ichishima, E. (1989) Inhibition study of tyrosinase from *Aspergillus oryzae. Agric. Biol. Chem.*, *53*, 557-558.

[149] Fujita, Y., Uraga, Y. and Ichisima, E. (1995). Molecular cloning and nucleotide sequence of the protyrosinase gene, *melO*, from *Aspergillus oryzae* and expression of the gene in yeast cells. *Biochim. Biophys. Acta*, *1261*, 151-154.

[150] Kuper, U., Niedermann, D. M., Travaglini, G. and Lerch, K. (1989). Isolation and characterization of the tyrosinase gene from *Neurospora crassa*. *J. Biol. Chem.*, *264*, 17250-17258.

[151] Huber, M., Hintermann, G., and Lerch, K. (1985). Primary structure of tyrosinase from *Streptomyces glaucescence*. *Biochemistry*, *24*, 6038-6044.

[152] Huber, M. and Lerch, K. (1988). Identification of two histidines as copper ligands in *Streptomyces glaucescens* tyrosinase. *Biochemistry*, *27*, 5610-5615.

[153] Known, B. S., Haq, A. K., Pomerantz, S. H. and Halaban, R. (1987). Isolation and sequence of cDNA for human tyrosinase that maps at the mouse-c-albino locus. *Proc. Natl. Acad. Sci. USA.*, *84*, 7473-7477.

[154] Lerch, K. (1982). Primary structure of tyrosinase from *Neurospora crassa*. II. Complete amino acid sequence and chemical structure of a tripeptide containing an unusual thioether. *J. Biol. Chem.*, *257*, 6414-6419.

[155] van Gelder, C. W., Flurkey, W. H. and Wichers, H. J. (1997). Sequence and structural features of plant and fungal tyrosinases. *Phytochemistry*, *45*, 1309-1323.

[156] Seo, S. Y., Sharma, V. K. and Sharma, N. (2003). Mushroom tyrosinase: recent prospects. *J. Agric. Food Chem.*, *51*, 2837-2853.

[157] Espín, J. C., van Leeuwen, J. and Wichers, H. J. (1999). Kinetic study of the activation process of a latent mushroom (*Agaricus bisporus*) tyrosinase by serine protease. *J. Agric. Food Chem.*, *47*, 3509-3517.

[158] Espín, J. C. and Wichers, H. J. (1999). Activation of a latent mushroom (*Agaricus bisporus*) tyrosinase isoform by sodium dodecyl sulfate (SDS). Kinetics properties of the SDS-activated isoform. *J. Agric. Food Chem.*, *47*, 3518-3525.

[159] Wittbjer, A., Odh, G., Rosengren, A. M., Rosengren, E. and Rorsman, H. (1989). Isolation of soluble tyrosinase from human melanoma cells. *Acta Derm. Venereol. 69*, 125-131.

[160] Wittenberg, C., and Triplett, E. L. (1985). A detergent-activated tyrosinase from *Xenopus laevis*. I. Purification and partial characterization. *J. Biol. Chem.*, *260*, 12535-12541.

[161] Penafiel, R., Galindo, J. D., Pedreno, E. and Lozano, J. A. (1982). The process for the activation of frog epidermis pro-tyrosinase. *Biochem. J.*, *205*, 397-404.

[162] Tatara, Y., Namba, T., Yamagata, Y., Yoshida, T., Uchida, T. and Ichishima, E. (2008). Acid activation of protyrosinase from *Aspergillus oryzae*: homo-tetrameric protyrosinase is converted to active dimmers with an essential intersubunit disulfide bond at acidic pH. *Pigment Cell Melanoma Res.*, *21*, 89-96.

[163] Slavik, J. (1982). Anilinonaphthalene sulfonate as a probe of membrane composition and function. *Biochim. Biophys. Acta*, *694*, 1-25.

[164] Klabunde, T., Eicken, C., Sacchettini, J. C. and Krebs, B. (1998). Crystal structure of a plant catechol oxidase containing a dicopper center. *Natl. Struct. Biol.*, *5*, 1084-1090.

[165] Imai, K. and Toyo'oka, T. (1987). Fluorometric assay of thiols with fluorobenzoxadiazoles. In Jakoby, W. (Ed.). *Methods in ENZYMOLOGY: SULFUR AND SULFUR AMINO ACIDS. Vol. 143*, (pp.67-75). New York: ACADEMIC PRESS.

[166] Blumental, C. Z. (2004). Production of toxic metabolites in *Aspergillus niger*, *Aspergillus oryzae*, and *Trichoderma reesei*: justification of mycotoxin testing in food grade enzyme preparations derived from the three fungi. *Regul. Toxicol. Pharmacol.*, *39*, 214-228.

[167] Obata, H., Ishida, H., Hata, Y., Kawato, A., Abe, Y., Akao, T. and Ichishima, E. (2004) Cloning of a novel tyrosinase-encoding gene (*melB*) from *Aspergillus oryzae* and its overexpression in solid-state culture (rice-koji). *J. Biosci. Bioeng.*, 97, 400-405.

[168] Narahara, H., Koyama, Y., Yoshida, T., Pichangkura, S., Ueda, R. and Taguchi, H. (1982). Growth and enzyme production in a solid-state culture of *Aspergillus oryzae. J. Ferment. Technol.*, *60*, 311-319.

[169] Hata, Y., Ishida, H., Kojima, Y., Ichikawa, E., Kawato, A., Suginami, K. and Imayassu, S. (1997). Comparison of two glucoamylases produced by *Aspergillus oryzae* in solid-state culture (koji) and in submerged culture. *J. Ferment. Bioeng.*, *84*, 532-537.

[170] Hata, Y., Ishida, H., Ichikawa, E., Kawato, A., Suginami, K. and Imayasu, S. (1998). Nucleotide sequence of an alternative glucoamylase-encoding gene (*glaB*) expressed in solid-state culture of *Aspergillus oryzae. Gene*, *207*, 127-132.

[171] Hata, Y., Tsuchiya, K., Kitamoto, K., Gomi,K., Kumagai, C., Tamura, G., and Hara, S. (1991). Nucleotide sequence and expression of the glucoamylase-encoding gene (*glaA*) from *Aspergillus oryzae. Gene*, *108*, 145-150.

[172] Minetoki, T., Nunokawa, Y., Gomi, K., Kitamoto, K., Kumagai, C. and Tamura, G. (1996) Deletion analysis of promoter elements of the *Aspergillus oryzae agdA* gene encoding α-glucosidase. *Curr. Genet.*, *30*, 432-438.

[173] Minetoki, T., Kumagai, C., Gomi, K., Kitamoto, K. and Takahashi, K. (1998). Improvement of promoter activity by the introduction of multiple copies of the conserved region III sequence, involved in the efficient expression of *Aspergillus oryzae* amylase-encoding gene. *Appl. Microbiol. Biotechnol.*, *50*, 459-467.

[174] Ishida, H., Matsumura, K., Hata, Y., Kawado, A., Siginami, K., Abe, Y., Imayasu, S. and Ichishima, E. (2001). Establishment of hyper-protein pruduction system in submerged *Aspergillus oryzae* culture under tyrosinase-encoding gene (*melO*) promoter control. *Appl. Microbiol. Biotechnol.*, *57*, 131-137.

INDEX

G

H

I

J

K

L

Q

R

S

T

U

V

W

Y

Z